PROTEIN SCIENCE AND ENGINEERING

PROTEINS RESEARCHER BIOGRAPHICAL SKETCHES AND RESEARCH SUMMARIES

PROTEIN SCIENCE AND ENGINEERING

Additional books in this series can be found on Nova's website
under the Series tab.

Additional E-books in this series can be found on Nova's website
under the e-books tab.

PROTEIN BIOCHEMISTRY, SYNTHESIS, STRUCTURE AND CELLULAR FUNCTIONS

Additional books in this series can be found on Nova's website
under the Series tab.

Additional E-books in this series can be found on Nova's website
under the e-books tab.

PROTEIN SCIENCE AND ENGINEERING

PROTEINS RESEARCHER BIOGRAPHICAL SKETCHES AND RESEARCH SUMMARIES

HUI-ZHONG WANG
AND
MIAO TIAN
EDITORS

Nova Science Publishers, Inc.
New York

For permission to use material from this book please contact us:
Telephone 631-231-7269; Fax 631-231-8175
Web Site: http://www.novapublishers.com

Library of Congress Cataloging-in-Publication Data

Proteins researcher biographical sketches and research summaries / editors, Hui-Zhong Wang and Miao Tian.
p. ; cm.
Includes bibliographical references and index.
ISBN 978-1-62100-777-7 (softcover)
I. Wang, Hui-Zhong. II. Tian, Miao, 1969-
[DNLM: 1. Biomedical Research--Biobibliography. 2. Proteins--Biobibliography. QD 21]
572'.633--dc23
2011038563

Published by Nova Science Publishers, Inc. † New York

Contents

PREFACE

This new book compiles biographical sketches of top professionals in the field of proteins, as well as research summaries from a number of different focuses in this important field.

PART 1 - RESEARCH BIOGRAPHIES

In: Proteins Researcher Biographical Sketches … ISBN: 978-1-62100-777-7
Editors: H. Z. Wang and M. Tian

Chapter 1

BIOGRAPHICAL SKETCH

NAME:	TITLE:	DATE OF BIRTH:
Nora Beatriz Calcaterra	IBR-CONICET; FAC. CS. BIOQUIMICAS Y FARMACEUTICAS - UNIVERSIDAD NACIONAL DE ROSARIO (UNR)	11/08/1959

EDUCATION:

Institution and Location	Degree	Year Conferred	Scientific Field
Universidad Nacional de Rosario	Biochemist	1978 - 1983	Biochemistry
Universidad Nacional de Rosario	Doctor (PhD)	1984 - 1988	Biochemistry
Universidad Nacional de Rosario	Postdoctoral Training	1988 - 1992	Molecular Biology
Centro de Biología Molecular "Severo Ochoa"; Univ. Autónoma de Madrid, Faculty of Sciences, Madrid, Spain	Visiting Professor	1997	Molecular Biology

CONTACT POINTS:

Address: Suipacha 590 S2002LRK - Rosario, Argentina

RESEARCH AND PROFESSIONAL EXPERIENCE:

2008 –continue
Faculty of Biochemistry and Pharmaceutical Sciences, UNR
Chairwoman of Biology Area
1999/2001
Faculty of Biochemistry and Pharmaceutical Sciences, UNR
Head of the Biotechnology Career.
1998-continue
National Research Council – IBR, Developmental Biology Division

Research Independent

1992/1998
National Research Council – IBR, Molecular Biology Division
Research Assistant
1988/1992
National Research Council – IBR, Molecular Biology Division
Post-Doctoral Fellow
1990/1995
Faculty of Biochemistry and Pharmaceutical Sciences. Biological Sciences Department – Molecular Biology Division.
Teaching Assistant
1984/1988
National Research Council– Centro de Estudios Fotosintéticos y Bioquimicos (CEFOBI)
Doctoral Fellow
1985/1990
Faculty of Biochemistry and Pharmaceutical Sciences. Biochemistry Sciences Department – Biochemistry Division.
Teaching Assistant

Professional Appointments:

Gene expression regulation during embryonic development. Role of single-stranded nucleic acid binding proteins.

Publications during Last Three Years:

1.18. Lombardo, V., Armas, P., Weiner, A.M.J., Calcaterra, NB (2007) "*In vitro* embryonic developmental phosphorylation of the cellular nucleic acid binding protein by cAMP-dependent protein kinase and its relevance for biochemical activities". *FEBS J.* 247: 485-497.

1.19. Weiner, A.; Allende, M, Becker, T.; Calcaterra, NB (2007) "CNBP mediates neural crest cell expansion by controlling cell proliferation and cell survival during rostral head development". *J. Cell. Biochem.*, 102:1553-1570.

1.20. Armas, P., Nasif, S., Calcaterra, NB (2008) "Cellular Nucleic acid Binding Protein binds G-rich single-stranded nucleic acids and may function as a nucleic acid chaperone". *J. Cell. Biochem.,* 103:1013-1036.

1.21. Armas, P. Agüero, T. Borgognone, M. Aybar, MJ; Calcaterra NB (2008) "Dissecting CNBP, a zinc-finger protein required for neural crest development, in its structural and functional domains" *J Mol Biol.*; 382:1043-56.

1.22. Weiner, AMJ; Allende, M; Calcaterra NB (2009) "Zebrafish *cnbp* intron1 plays a fundamental role in controlling spatiotemporal gene expression during embryonic development" *J. Cell. Biochem.,* 108:1364-1375

1.23. Mussi, MA and Calcaterra, NB (2010) "Paraquat-Induced Oxidative Stress Response During Amphibian Early Embryonic Development" *Comp Biochem Phys C,* 151:240-247

1.24. Borgognone M, Armas P and Calcaterra NB (2010) "Cellular nucleic acid binding protein, a transcriptional enhancer of c-Myc, promotes the formation of parallel G-quadruplexes" *Biochem J*, 428: 491- 498

1.25. Calcaterra, NB; Armas, P; Weiner, AMJ; Borgognone, M. (2010) "CNBP: a multifunctional nucleic acid chaperone involved in cell death and proliferation control", IUBMB *Life*, 62: 707-714.

In: Proteins Researcher Biographical Sketches …
Editors: H. Z. Wang and M. Tian
ISBN: 978-1-62100-777-7

Chapter 2

BIOGRAPHICAL SKETCH

NAME:	TITLE:	DATE OF BIRTH:
Richard A. Armstrong	VISION SCIENCES, ASTON UNIVERSITY	25.11.1949

EDUCATION:

Institution and Location	Degree	Year Conferred	Scientific Field
King's College University of London; University of Oxford (St Catherine's College)			

CONTACT POINTS:

Address: Vision Sciences, Aston University, Birmingham B4 7ET, UK

RESEARCH AND PROFESSIONAL EXPERIENCE:

Neuropathology of Neurodegenerative diseases, Plant ecology especially lichens

Professional Appointments:

Lecture in Neurobiology, Aston University 1976-2011

Publications during Last Three Years:

Armstrong RA (2009a) Pathological changes in the sporadic and variant forms of Creutzfeldt-Jakob disease (CJD) are spatially related to blood vessels. Brain Research Journal 2 (3) 149-162.

Armstrong RA (2009b) Monthly fluctuations in radial growth of individual lobes of the lichen Parmelia conspersa (Ehrh. Ex Ach.) Ach. Symbiosis 47: 9-16.

Armstrong RA (2009c) Spatial correlations between β-amyloid (Aβ) deposits and blood vessels in early-onset Alzheimer's disease. In: Alzheimer's Disease in the Middle-Aged. Ed: Hyun Sil Jeong, Nova Science Publishers, pp 137-153.

Armstrong RA (2009d) Spatial correlations between βamyloid (Aβ) deposits and blood vessels in familial Alzheimer's disease. Folia Neuropathol 46: 241-248.

Armstrong RA, Lantos PL, Cairns NJ (2009e) Hippocampal pathology in progressive supranuclear palsy (PS): a quantitative study of 8 cases. Clin Neuropathol 28: 46-53.

Armstrong RA, Ironside JW, Lantos PL, Cairns NJ (2009f) A quantitative study of the pathological changes in the cerebellum in 15 cases of variant Creutzfeldt-Jakob disease (vCJD). Neuropathol Appl Neurobiol 35: 36-45.

Armstrong RA (2009g) Relationship between the microcirculation and the pathological changes in the sporadic and variant forms of Creutzfeldt-Jakob disease (CJD). In:Microcirculation: Function, Malfunction and Measurement. Ed: Ralph Thompson, Nova Science Publishers, pp 79-94.

Armstrong RA, Cairns NJ (2009h) Laminar distribution of the pathological changes in frontal and temporal cortex in 8 patients with progressive supranuclear palsy. Clin Neuropathol 28: 350-357.

Armstrong RA, Cairns NJ (2009i) Clustering and spatial correlations of the neuronal cytoplasmic inclusions, astrocytic plaques and ballooned neurons in corticobasal degeneration. J Neural Transm. 116: 1103-1110.

Armstrong RA (2009j) Spatial correlations between the vacuolation, prion protein (PrPsc) deposits and the cerebral blood vessels in sporadic Creutzfeldt-Jakob disease. Curr Neurovasc Res 6: 239-245.

Armstrong RA, Cairns NJ (2009k) Size frequency distribution of the □-amyloid (Aβ) deposits in dementia with Lewy bodies with associated Alzheimer's disease pathology. Neurol Sci 30: 471-477.

Armstrong RA, Smith SN (2009l) Carbohydrates in the hypothallus and areolae of the crustose lichen Rhizocarpon geographicum (L.) DC. Symbiosis 49: 95-100.

Armstrong RA (2009m) Alzheimer's disease and the eye. J Optom 2: 103-111

Hurrell E, Kucerova E, Loughlin M, Caubilla-Barron J, Hilton A, Armstrong RA, Smith C, Grant J, Shoo S, Forsythe S (2009n) Neonatal enteral feeding tubes as loci for colonization by members of the Enterobacteriaceae. BMC Infectious Diseases 9: 146-155.

Armstrong RA (2009o) The molecular biology of senile plaques and neurofibrillary tangles in Alzheimer's disease. Folia Neuropathol 47: 289-299.

Armstrong RA (2010a) A quantitative study of the pathological changes in the cortical white matter in variant Creutzfeldt-Jakob disease (vCJD). In:Handbook on White Matter. Ed: TB Westland, RN Calton, Nova Science Publishers, pp 133-146.

Armstrong RA (2010b) The spatial patterns of tau immunopositive neuronal cytoplasmic inclusions (NCI) in the tauopathies. In:Neuroscience Research Advances. Ed: B. Figueredo, F Melandez, Nova Science Publishers, pp 175-189.

Armstrong RA, Ellis W, Hamilton RL, Mackenzie IRA, Hedreen J, Gearing M, montine T, Vonsattel JP, Nead E, Lieberman AP, Cairns NJ (2010c) Neuroapthological heterogeneity in frontotemporal lobar degeneration with TDP-43 proteinopathy: a quantitative study of 94 cases using principal components analysis. J Neural Transm 117: 227-239.

Armstrong RA, Bradwell T (2010d) Growth of crustose lichens: a review. Geogr Ann 92A: 3-17.

Armstrong RA, Bradwell T (2010e) The use of lichen growth rings in lichenometry: Some preliminary findings. Geogr Ann 92A: 141-147.

Armstrong RA, Cairns NJ (2010f) Analysis of β-amyloid (Aβ) deposition in the temporal lobe in Alzheimer's disease using Fourier (spectral) analysis. Neuropathol Appl Neurobiol 36: 248-257.

Armstrong RA (2010g) A spatial pattern analysis of β-amyloid (Aβ) deposition in the temporal lobe in Alzheimer's disease. Folia Neuropathol 48: 67-74.

Armstrong RA (2010h) Dispersion of prion protein deposits around blood vessels in variant Creutzfeldt-Jakob disease. Folia Neuropathol 48: 166-174.

Armstrong RA (2010i) Lobe formation and division in the foliose lichen Xanthoparmelia conspersa. Symbiosis 51: 227-232.

Armstrong RA (2010j) A quantitative study of the pathological changes in the cortical white matter in variant Creutzfeldt-Jakob diseases (vCJD). Clin Neuropathol 29: 390-396.

Smith SN, Khan JA, Armstrong RA, Mohindru B, Prince M, Whipps JM (2010k) Acanthamoeba polyphaga trophozoite binding of representative fungal single cell forms. Acta Protozoologica 49: 289-300.

Armstrong RA (2010l) Quantitative methods in neuropathology. Folia Neuropathol 48: 217-230.

Armstrong RA, Cairns NJ (2011a) A morphometric study of the spatial patterns of the TDP-43-immunoreactive neuronal inclusions in frontotemporal lobar degeneration with progranulin (GRN) mutation. Histol Histopathol 26: 185-190.

Armstrong RA (2011b) Laminar distribution of the pathological changes in sporadic and variant Creutzfeldt-Jakob disease. Pathology Research International. Volume 2011, Article ID 236346, 7 pages.

Armstrong Ra, Gearing M, Bigio EH, Cruz-Sanchez FF, Duyckaerts, Mckenzie IRA, Perry RH, Skullerud K, Yokoo H, Cairns NJ (2011c) The spectrum and severity of FUS-immunoreactive inclusions in the frontal and temporal lobes of ten cases of neuronal intermediate filament inclusion disease. Acta Neuropathol 121: 219-228.

In: Proteins Researcher Biographical Sketches ...
Editors: H. Z. Wang and M. Tian
ISBN: 978-1-62100-777-7

Chapter 3

BIOGRAPHICAL SKETCH

NAME:	TITLE:	DATE OF BIRTH:
Mira Barda-Saad	**PHD THE MINA AND EVERARD GOODMAN FACULTY OF LIFE SCIENCES, BAR-ILAN UNIVERSITY, RAMAT-GAN 52900, ISRAEL.**	

CONTACT POINTS:

Address: The Mina & Everard Goodman Faculty of Life Sciences, Bar-Ilan University, Ramat-Gan 52900, Israel

Email: bardasm@mail.biu.ac.il

RESEARCH AND PROFESSIONAL EXPERIENCE:

Mira Barda-Saad received her BSc and MSc degrees (*cum laude*) from Bar-Ilan University. Her PhD thesis and postdoctoral studies focused on the T-cell antigen receptor (TCR) structure and signal transduction. She received her PhD in 1999 (*summa cum laude*) and first pursued her postdoctoral studies at the Department of Molecular Cell Biology at the Weizmann Institute of Science (WIS), Rehovot (1999-2001). During her PhD and postdoctoral studies, she discovered that mesenchymal cells express components of the CD3/TCR complex relevant to cell growth control (*Oncogene* 21:2029, 2002). Two patents were filed based on these novel findings because of their relevance to basic science and their potential for clinical application, such as gene therapy in genetic or acquired hematopoietic disorders by manipulating mesenchymal cell growth and functions. These novel findings led to the discovery of immunoglobin expression in mesenchymal cells, which was consequently filed as an additional patent.

A second postdoctoral study at the National Institute of Health (NIH) focused on the signal transduction mechanisms in T lymphocyte activation. Barda-Saad explored the dynamics of the molecular interactions linking the TCR to cytoskeleton reorganization (*Nature Immunology* 6:80, 2005). This work was rated as a "Must-Read" publication by the "Faculty of 1000". It also prompted an invitation to contribute an article to the periodical of the NIH's National Cancer Institute, *Frontiers in Science*, which reviews major scientific advancements by the institute's scientists (*CCR Frontiers in Science*, 2006). In October 2006, she was appointed as a faculty member of the Faculty of Life Sciences at Bar-Ilan University, where she serves as an Assistant Professor. Her laboratory focuses on gaining a better understanding of the mechanisms that underlie immune cell signaling, with a primary goal of relating this knowledge to the physiology and pathophysiology of the immune response. The main topic currently investigated in her laboratory is the way signaling proteins regulate multiple pathways of cytoskeleton reorganization, specifically, actin polymerization. This is accomplished by utilizing a multidisciplinary approach including molecular imaging, biochemical and biophysical techniques. Her research goal is to discover the regulation of multimolecular complexes of signaling molecules critical for cytoskeletal reorganization and, indeed, her research work during the last few years has yielded several important contributions to the fields of cellular biology, immunology, nanobiology, and nanoclustering (*EMBO J.* 29:2315-2328, 2010, *Comm. Integ. Biol.* 4:175-177, 2011, and *Mol. Cell. Biol.* In press, 2011). Using high-resolution molecular imaging technology, her research group discovered the signaling cascade that regulates the recruitment of the cytoskeletal proteins to the immunological synapse and also the dynamic behavior of signaling and cytoskeletal molecules regulating immune cell activation and function.

Professional Appointments:

Assistant Professor, The Mina & Everard Goodman Faculty of Life Sciences, Bar-Ilan University, Ramat-Gan, Israel

Honors:

Dr. Barda-Saad has received numerous awards, including the Leon and Maria Taubenblatt Award for Excellence in Medical Research (2009) and an award for excellence in research from the Israeli Cancer Association (2008). She has also been awarded the FARE Award (Fellow Award for Research Excellence) at NIH (2004) and a special award for excellence from the Israeli Knesset (parliament).

Publications during Last Three Years:

1) Barda-Saad, M., Braiman, A., Titerence, R., Bunnell, S.C., Barr, V.A., and Samelson, L.E. (2005) Dynamic molecular interactions linking the T cell antigen receptor to the actin cytoskeleton. *Nature Immunology* 6:80-89.

2) Houtman, J., Houghtling, R., Barda-Saad, M., Toda, Y., Schwartzberg, PL., and Samelson, LE. (2005). Early phosphorylation kinetics of proteins involved in proximal T-cell receptor-mediated signaling pathways. *J. Immunol.* 15:2449-2458.
3) Houtman, J., Barda-Saad, M. and Samelson, L.E. (2005). How to examine multiprotein signaling complexes from all angles: The use of complementary techniques to characterize complex formation at the adapter protein LAT. *FEBS J.* 272:5426-35.
4) Barda-Saad, M., Braiman, A., Titerence, R., Barr, V. and Samelson, L.E. (2006). T cell antigen receptor engagement induced cytoskeleton re-organization. *CCR Frontiers in Science* (Published by the National Cancer Institute's Center for Cancer Research, NIH).
5) Braiman, A., Barda-Saad, M. and Samelson, L.E. (2006). Recruitment and activation of PLCγ1in T-cells: a new insight into old domains. *EMBO J.* 25:774-84.
6) Barr, V.A., Balagopalan-Bhise, L., Barda-Saad, M., Polishchuk, R., Boukari, H., Bunnell, S.C., Bernot, K.M., Toda, Y., Nossal, R., and Samelson, L.E. (2006). T-cell antigen receptor-induced signaling complexes: Internalization via a cholesterol-dependent endocytic pathway. *Traffic* 7:1143-1163.
7) Houtman, J., Yamaguchi, H., Barda-Saad, M., Bowden, B., Schuck, P., Appella, E., and Samelson, LE. (2006). Oligomerization of intracellular signaling complexes by the multipoint binding of the adapter protein Grb-2 to both LAT and Sos1. *Nature Struct. Mol. Biol.* 13:798-805.
8) Balagopalan, L., Barr, V.A., Sommers, C.L., Barda-Saad, M., Goyal, A., Isakowitz, M.S. and Samelson, L.E. (2007). c-Cbl mediated regulation of LAT-nucleated signaling complexes. *Mol. Cell. Biol.* 27:8622-8636.
9) Sommers, C., Gurson, J.M., Surana, R., Barda-Saad, M., Lee, J., Kishor, A., Li, W., Gasser, A.J., Barr, V., Miyaji, M., Love, P. and Samelson, L.E. (2008) Bam-32: a novel mediator of Erk activation in T cells. *Int. Immunol.* 20:811-818.
10) Shani N, Rubin-Lifshitz, H., Peretz-Cohen Y. Shkolnik, K., Shnider, V., Cohen-Sfady, M., Shav-Tal, Y., Barda-Saad, M. and Zipori D. (2009) Incomplete T cell receptor transcripts encode proteins that target the mitochondrion and induce apoptosis. *Blood* 9:3530-3541.
11) Barda-Saad, M.*, Shirasu, N., Pauker, M.H., Hasan, N., Perl, O., Balbo, A., Yamaguchi, H., Houtman, J., Appella, E., Schuck, P. and Samelson, L.E. (2010) Cooperative interactions at the SLP-76 complex are critical for T-cell signaling and actin polymerization. *EMBO J.* 29:2315-2328. **Corresponding author*
12) Reicher, B. and Barda-Saad, M.* (2010) Multiple pathways leading from the T-cell antigen receptor to the actin cytoskeleton network. *FEBS Lett.* 584:4858-4864. **Corresponding author*
13) Hazani, M. and Barda-Saad, M.* (2010) Studies of novel interactions between Nck and VAV SH3 domains. *Comm. Integ. Biol.* 4:175-177. **Corresponding author.*
14) Pauker, H.M., Reicher, B., Fried, S., Perl, O., and Barda-Saad, M.* (2011) Functional cooperativity between the proteins Nck and ADAP is fundamental for actin reorganization following TCR activation. *Mol. Cell. Biol. In press. *Corresponding author*

ISBN: 978-1-62100-777-7

Chapter 4

BIOGRAPHICAL SKETCH

NAME:	TITLE:	DATE OF BIRTH:
Miles F. Beaux II	**LOS ALAMOS NATIONAL LABORATORY**	**10 JUNE 1979**

EDUCATION:

Institution and Location	Degree	Year Conferred	Scientific Field
University of Idaho	PhD Physics	2010	
University of Idaho	MS Physics	2008	
University of Idaho	BS Physics and Idaho State Teaching Certificate	2005	
Brigham Young University	Teaching Certification, Teacher of English as a Foreign Language	1999	

CONTACT POINTS:

Address: MS K764
Los Alamos Natl. Lab
P.O. Box 1663; Los Alamos, NM 87545

RESEARCH AND PROFESSIONAL EXPERIENCE:

Actinide photoemission research: measurements of d-Pu and Pu oxide at LANL, Synchrotron Radiation Center measurement experience of U materials.

Project lead for bioapplications of nanospring™ materials, nanomaterial coatings expert: developed metal and metal oxide coating processes for nanospring™ material via atomic layer deposition, chemical vapor deposition, and dip coat methods.

Biological applications of nanotechnologies: drug delivery and nanotoxicology, one dimensional nanomaterial fabrication via vapor-liquid-solid mechanism, surface analysis of nanomaterials and their interactions with biomolecules (i.e. fibronectin, shiga toxin, green fluorescing protein, etc.) via X-ray photoelectron, FTIR, and Raman spectroscopy.

Professional Appointments:

Los Alamos National Laboratory, Postdoc, Sept. 2011 – present.
GoNano Technologies, Inc., Nanomaterials Engineer, May 2008 – Aug. 2011.
University of Idaho, Research Assistant/Nanomaterials Researcher.

Honors:

Research Assistant of the Year Award. University of Idaho Physics Department, 2009-2010.

Glen E. and Jean K. Nielsen Science Scholarship, 2007-2008.

Publications during Last Three Years:

Hard X-Ray Photoelectron Spectroscopy and Electronic Structure of single crystal UPd_3, UGe_2, and USb_2. Submitted to *Journal of Physics – Condensed Matter* (2011), in review.

Reynolds number manipulation of mean nanowire dimensions and nanowire suspension quantification. Submitted to *Journal of Applied Physics* (2011), in review.

X-Ray Photoelectron Spectroscopic Analysis of the Surface Chemistry of Silica Nanowires. *Applied Surface Science* 257, 5766-71 (2011).

Toxic and teratogenic silica nanowires in developing vertebrate embryos. *Nanomedicine: Nanotechnology, Biology and Medicine* 6, 93-102 (2010).

In vitro proliferating cell models to study cytotoxicity of silica nanowires. *Nanomedicine: Nanotechnology, Biology and Medicine* 6, 84-92 (2010).

In: Proteins Researcher Biographical Sketches … ISBN: 978-1-62100-777-7
Editors: H. Z. Wang and M. Tian

Chapter 5

BIOGRAPHICAL SKETCH

NAME:	TITLE:	DATE OF BIRTH:
Miodrag Belosevic	DR. UNIVERSITY OF ALBERTA	

EDUCATION:

Institution and Location	Degree	Year Conferred	Scientific Field
University of Manitoba	B.Sc.	1976	
University of Manitoba	M.Sc.	1979	
McGill University	Ph.D. (with distinction	1985	
McGill Centre for Tropical Medicine, Montreal	PDF	1985-1986	
Walter Reed Army Institute of Research, Washington DC	PDF	1986-1988	

CONTACT POINTS:

Address: CW-405 Biological Sciences, University of Alberta, Edmonton, Alberta, Canada T6G 2E9

RESEARCH AND PROFESSIONAL EXPERIENCE:

Present Position: Distinguished University Professor of Biological Sciences and Public Health Sciences and Stream Leader, Environmental Health Sciences, University of Alberta.

Research: Main research interests include elucidation of mechanisms of host defense in protozoan infections, innate immunity in fish and control of waterborne infectious diseases.

Publications: >270 refereed publications, >100 research presentations and seminars, >200 conference presentations.

Honors:

Recipient of more than 25 awards including Clark P. Read Mentor Award, American Society of Parasitologists (2009); Thomas C. Keefer Medal, Canadian Society for Civil Engineering (2009); Fellow of the Royal Society of Canada (2008); Distinguished University Professor, University of Alberta (2008); University Cup, University of Alberta (2006); Alberta Centennial Medal, Province of Alberta (2005); Killam Award for Excellence in Mentoring, Killam Trust (2004); Alberta Science and Technology Foundation Award (ASTech) (2003); Wardle Medal, Canadian Society of Zoologists (2002); Rudolph Hering Medal, American Society of Civil Engineers (2002); Killam Annual Professorship, Killam Trust (2001).

<u>Publications during Last Three Years:</u>

(Select publications; Total 56)

Grayfer L, Hodgkinson JW, Hitchen SJ, Belosevic M. Characterization and functional analysis of goldfish (*Carasius auratus* L.) interleukin-10. *Molecular Immunology* 2011; 48:563-571.

Grayfer L. Belosevic M. Comparison of macrophage antimicrobial responses induced by type II interferons of the goldfish (*Carasius auratus* L.). *Journal of Biological Chemistry* 2010; 285:23537-23547.

Oladiran A, Belosevic M. *Trypanosoma carassii* calreticulin binds host complement component C1q and inhibits classical complement pathway-mediated lysis. *Developmental and Comparative Immunology* 2010; 34:396-405.

Grayfer L, Belosevic M. Molecular characterization of novel interferon gamma receptor 1 isoforms in zebrafish (*Danio rerio*) and goldfish (*Carassius auratus* L.). *Molecular Immunology* 2009; 46:3050-3059.

Katzenback BA, Belosevic M. Molecular and functional characterization of *kita* and *kita* of the goldfish (*Carassius auratus* L.). *Developmental and Comparative Immunology* 2009; 33: 1165-1175.

Hitchen SJ, Shostak AW, Belosevic M. *Hymenolepis diminuta* (Cestoda) induces changes expression of select genes of *Tribolium confusum* (Coleoptera). *Parasitology Research* 2009; 105:875-879.

Oladiran A, Belosevic M. *Trypanosoma carassii* hsp70 increases the expression of inflammatory cytokines and chemokines in macrophages of the goldfish (*Carassius auratus* L.). *Developmental and Comparative Immunology* 2009; 33: 1128-1136.

Li D, Craik SA, Smith DW, Belosevic M. Infectivity of *Giardia lamblia* cysts obtained from wastewater treated with ultraviolet light. *Water Research* 2009; 43:3037-3046.

Grayfer L, Belosevic M. Molecular characterization of tumor necrosis factor receptor 1 and 2 of the goldfish (*Carassius auratus* L.). *Molecular Immunology* 2009; 46:2190-2199.

Grayfer L, Hanington PC, Belosevic M. Macrophage colony-stimulating factor (CSF-1) induces pro-inflammatory gene expression and enhances antimicrobial responses of goldfish (*Carassius auratus* L.) macrophages. *Fish and Shellfish Immunology* 2009; 26:406-413.

Hanington PC, Hitchen SJ, Beamish LA, Belosevic M. Macrophage colony stimulating factor (CSF-1) is a central growth factor of goldfish macrophages. *Fish and Shellfish Immunology* 2009; 26:1-9.

Hanington PC, Tam J, Katzenback BA, Hitchen SJ, Barreda DR, Belosevic M. Development of macrophages of bony fish. *Developmental and Comparative Immunology* 2009; 33:411-429.

Katzenback BA, Belosevic M. Isolation and functional characterization of neutrophils from goldfish (*Carassius auratus* L.) kidney. *Developmental and Comparative Immunology* 2009; 33:601-611.

Li D, Craik SA, Smith DW, Belosevic M. The assessment of particle association and UV disinfection of wastewater using indigenous spore-forming bacteria. *Water Research* 2009; 43:481-489.

Haddad G, Belosevic M. Transferrin-derived synthetic peptide induces highly conserved pro-inflammatory responses of macrophages. *Molecular Immunology* 2009; 46:576-586.

Grayfer L, Belosevic M. Molecular characterization, expression and functional analysis of goldfish (*Carassius auratus* L.) interferon-gamma. *Developmental and Comparative Immunology* 2009; 33:235-246.

Wang T, Hanington PC, Belosevic M, Secombes CJ. Fish macrophage colony stimulating factor genes: diversified gene organizations and novel functions. *Journal of Immunology* 2008; 181:3310-3322.

Oladiran A, Hitchen SJ, Katzeback BA, Belosevic M. Biology of select zoonotic protozoan infections of domestic animals. UNESCO *Encyclopedia of Life Support Systems* 2008; (http://greenplanet.eolss.net/EolssLogn/InsLogin.aspx)

Kerr JL, Guo Z, Smith DW, Goss GG, Belosevic M. Use of goldfish to monitor wastewater and reuse water for xenobiotics. *Journal of Environmental Engineering and Science* 2008; 7: 369-383.

Hanington PC, Brennan LJ, Belosevic M, Keddie BA. Molecular and functional characterization of granulin-like molecule of insects. *Insect Biochemistry and Molecular Biology* 2008; 38:596-603.

Haddad G, Hanington PC, Wilson EC, Grayfer L, Belosevic M. Molecular and functional characterization of goldfish (*Carassius auratus* L.) transforming growth factor-beta. *Developmental and Comparative Immunology* 2008; 32:654-663.

Hanington PC, Patten SA, Reaume LM, Waskewitz AJ, Belosevic M, Ali DW. Analysis of leukemia inhibitory factor and leukemia inhibitory factor receptor in embryonic and adult zebrafish (*Danio rerio*). *Developmental Biology* 2008; 314:250-260.

Grayfer L, Walsh JG, Belosevic M. Characterization and functional analysis of goldfish (*Carassius auratus* L.) tumor necrosis factor-alpha. *Developmental and Comparative Immunology* 2008; 32:532-543.

Katzenback BA, Plouffe DA, Haddad G, Belosevic M. Immunization with recombinant parasite beta-tubulin increases resistance of goldfish *Carassius auratus* (L.) to infection with *Trypanosoma danilewskyi* Laveran and Mesnil, 1904. *Veterinary Parasitology* 2008; 151:36-45.

Li D, Craik SA, Smith DW, Belosevic M. Survival of *Giardia lamblia* trophozoites after exposure to ultraviolet light. *FEMS Microbiology Letters* 2008; 278:56-61.

Hanington PC, Wang T, Secombes CJ, Belosevic M. Growth factors of lower vertebrates: Characterization of goldfish (*Carassius auratus* L.) macrophage colony stimulating factor-1. *Journal of Biological Chemistry* 2007; 282:31865-31872.

Li D, Craik SA, Smith DW, Belosevic M. Comparison of inactivation of two isolates of *Giardia lamblia* by UV light. *Applied and Environmental Microbiology* 2007; 73:2218-2223.

Walsh JG, Barreda DR, Belosevic M. Cloning and expression analysis of goldfish (*Carassius auratus* L.) prominin. *Fish and Shellfish Immunology* 2007; 22:308-317.

In: Proteins Researcher Biographical Sketches ... ISBN: 978-1-62100-777-7
Editors: H. Z. Wang and M. Tian

Chapter 6

BIOGRAPHICAL SKETCH

NAME:	TITLE:	DATE OF BIRTH:
Sean J. Blamires	TUNGHAI UNIVERSITY	10 FEBRUARY 1972

EDUCATION:

Institution and Location	Degree	Year Conferred	Scientific Field
University of Sydney, Australia	PhD.	2007	
University of New South Wales, Australia	Grad. Dip. Ed.	2003	
Northern Territory University, Australia	MSc.	2000	
Deakin University, Australia	BSc.	1995	

CONTACT POINTS:

Address: Department of Life Science, Tunghai University. No 181 Section 3, Taichung-kan Road, Taichung city, Taiwan 407-04, R.O.C.

RESEARCH AND PROFESSIONAL EXPERIENCE:

2009 onward: Postdoc research on spider silk and web plasticity at Tunghai University, Taiwan, advised by I-Min Tso.

2004-2007: PhD on spider web and behavioural plasticity at University of Sydney, Australia, supervised by Michael Thompson and Dieter Hochuli.

1997-2000: MSc research on sea turtle and monitor lizard ecology at Northern Territory University, Australia, supervised by Michael Guinea. Published 20+ papers, 10 on spider webs and silk (as of Jan 2011).

Professional Appointments:

2009- Postdoctoral Research Fellow, Department of Life Science, Tunghai University, Taiwan. 2008: Lecturer, University of Technology, Sydney.

Honors:

1997: Northern Territory University.

Publications during Last Three Years:

1. Blamires, S.J., Chao, I.C., Liao, C.P. & Tso, I.M. 2011. Multiple prey cues combine to induce foraging flexibility in a trap-building predator. Animal Behaviour. In press.
2. Blamires, S.J. 2011. Nutritional implications for sexual cannibalism in a sexually dimorphic orb web spider. Austral Ecology. 35: In press.
3. Blamires, S.J. & Tso, I.M. 2011. Genetic –environmental influences over spider silk performance: a cross-clade comparison 0f orb-web spiders. Chinese Bioscience 53: 48.
4. Blamires, S.J., Lee, Y.H., Chang, C.M., Lin, I.T., Chen, J.A., Lin, T.Y. & Tso, I.M. 2010. Multiple structures interactively influence prey capture efficiency in spider orb webs. Animal Behaviour 80: 947-953.
5. Blamires, S.J., Chao, I.C. & Tso, I.M. 2010. Prey type, vibrations and handling interactively influence spider silk expression. Journal of Experimental Biology 213: 3906-3910.
6. Blamires, S.J. 2010. Plasticity in extended phenotypes: orb spider web architectural responses to variations in prey parameters. Journal of Experimental Biology 213: 3207-3212.
7. Blamires, S.J., Hochuli, D.F. & Thompson, M.B. 2009. Prey protein content influences growth and web decoration building in an orb building spider. Ecological Entomology 34: 545-550.
8. Blamires, S.J., Hochuli, D.F. & Thompson, M.B. 2008. Why the spider crossed the web: affects of decoration spectral properties on the number and type of natural prey captured in an orb spider (*Argiope keyserlingi*) web. Biological Journal of the Linnean Society. 94: 221-229.

In: Proteins Researcher Biographical Sketches … ISBN: 978-1-62100-777-7
Editors: H. Z. Wang and M. Tian

Chapter 7

BIOGRAPHICAL SKETCH

NAME:	TITLE:	DATE OF BIRTH:
Vilceu Bordignon	**DEPARTMENT OF ANIMAL SCIENCE, MCGILL UNIVERSITY, STE-ANNE-DE-BELLEVUE, QUEBEC, CANADA.**	

EDUCATION:

Institution and Location	Degree	Year Conferred	Scientific Field
Faculty of veterinary medicine, URCAMP, Brazil.	DVM	1984-1988	
Faculty of Veterinary Medicine, Federal University of Pelotas, Brazil.	M.Sc. in Veterinary sciences	1990-1993	
Faculté de Médecine Vétérinaire, Université de Montréal	Ph.D. in Biomedical sciences	1995-1999	
Centre de Recherche en Reproduction Animale, Faculté de Médecine Vétérinaire, Université de Montréal.	Postdoctoral training	1999-2002	

CONTACT POINTS:

Address: Department of Animal Science, McGill University

21,111 Lakeshore, Ste-Anne-de-Bellevue, Quebec
H9X 3V9 – Canada

Professional Appointments:

1989-1990: Veterinary Residency Program: Faculty of veterinary medicine, URCAMP, Bagé – RS, Brazil.
1991-1994: Veterinary practitioner in animal reproduction: SuperGen Animal Genetics, Capão do Leão - RS, Brazil.
1999-2002: Postdoctoral fellow: Faculté de Médecine Vétérinaire, Université de Montréal.
2002-2008: Assistant Professor, Department of Animal Science, McGill University, Canada.
2002-now: Associate Professor, Department of Animal Science, McGill University, Canada.

Honors:

1998. Runner-up award, "Student research presentation" at the meeting of the International Embryo Transfer Society in Boston, Massachusetts, USA in January 1998. Presentation title: "Effect of cytoplast cell cycle stage on the reprogramming of somatic histone H1 in reconstructed bovine embryos".
1999. Dean's honor roll nomination. Awarded to the top 5% of graduate students at graduation. Faculty of Graduate Studies, Biomedical Sciences PhD program, Université de Montréal.
2004. New Opportunities Fund Award. Canadian Foundation for Innovation.

Publications during Last Three Years:

Total of **20** scientific publications for the last 3 years.

2011

1. Baldassarre H, Deslauriers J, Neveu N and Bordignon V, 2011. Detection of endoplasmic reticulum stress markers and production enhancement treatments in transgenic goats expressing recombinant human butyrylcholinesterase. Transgenic Research, 2011 Feb 22. [Epub ahead of print]
2. Lisboa LA, Bordignon V and Seneda MM, 2011. Immunolocalization of brg1 –swi/snf protein during folliculogenesis in the porcine ovary. Zygote, 2011 Apr 15:1-6. [Epub ahead of print]
3. Coutinho ARS, Assumpção MEO and Bordignon V, 2011. Presence of active caspase 3 and developmental capacity of in vitro produced swine embryos. Molecular Reproduction and Development (accepted).

2010

4. Brochu-Gaudreau K, Rehfeldt C, Blouin R, Bordignon V, Murphy BD and Palin M-F, 2010. Adiponectin Action from Head to Toe. Endocrine 37:11–32.
5. Martinez-Diaz MA, Che L, Albornoz M, Seneda MM, Collis D, Coutinho ARS, El-Beirouthi N, LaurinD, Zhao X, Bordignon V, 2010. Pre- and Post-Implantation Development of Swine Cloned Embryos Derived from Fibroblasts and Bone Marrow Cells after Inhibition of Histone Deacetylases. Cellular Reprogramming (Cloning Stem Cells) 12:85-94.
6. Houde AA, Méthot S, Murphy BD, Bordignon V and Palin MF, 2010. Relationships between backfat thickness and reproductive efficiency of sows: a two year trial on two commercial herds fixing their backfat thickness at breeding. Canadian Journal of Animal Science 90:429-436.
7. Nascimento AB, Albornoz MS, Che L, Visintin JA and Bordignon V, 2010. Synergistic effect of porcine follicular fluid and dibutyryl cyclic adenosine monophosphate on development of parthenogenetically activated oocytes from pre-pubertal gilts. Reproduction in Domestic Animals 45:851–859.

2009

8. Labrecque B, Mathieu O, Bordignon V, Murphy BD and Palin NF, 2009. Identification of differentially expressed genes in a porcine in vivo model of adipogenesis using suppression subtractive hybridization. Comparative Biochemistry and Physiology Part D: Genomics and Proteomics 4:32-44.
9. Zhang W, Poirier L, Martinez-Diaz M, Bordignon V and Clarke HJ, 2009. Maternally encoded stem-loop-binding protein is degraded in 2-cell mouse embryos by the co-ordinated activity of independently regulated pathways. Developmental Biology 328:140-147.
10. Baldassarre H, Schirm M, Deslauriers J, Turcotte C and Bordignon V, 2009. Protein profile and alpha-lactalbumin concentration in the milk of standard and transgenic goats expressing recombinant human butyrylcholinesterase. Transgenic Research 18:621-632.
11. Labrecque B, Beaudry D, Mayhue M, Hallé C, Bordignon V, Murphy BD and Palin NF, 2009. Molecular characterization and expression analysis of the porcine paraoxonase 3 (PON3) gene. Gene 443:110–120.

2008

12. Bastos GM, Gonçalves PBD, Bordignon V, 2008. Immunolocalization of the high-mobility group N2 protein and acetylated histone H3K14 in early developing parthenogenetic bovine embryos derived from oocytes of high and low developmental competence. Molecular Reproduction and Development 75:282-90.
13. Campos DB, Palin M-F, Bordignon V, Murphy BD, 2008. The “Beneficial” Adipokines in Reproduction and Fertility. International Journal of Obesity, 32:223-31.
14. Baldassarre H, Hockley DK, Doré M, Brochu E, Hakier B, Zhao X, Bordignon V, 2008. Lactation performance of transgenic goats expressing recombinant human butyryl-cholinesterase in the milk. Transgenic research 17:73-84.
15. Houde AA, Murphy BD, Mathieu O, Bordignon V, Palin MF, 2008. Characterization of swine adiponectin and adiponectin receptors polymorphisms and their association with reproductive traits. Animal Genetics 39:249-57.

16. Seneda MM, Godmann M, Murphy BD, Kimmins S, Bordignon V, 2008. Developmental regulation of histone H3 methylation at lysine 4 in the porcine ovary. Reproduction 135:829-38.
17. Palin MF, Labrecque B, Beaudry D, Bordignon V, Murphy BD, 2008. Increased expression of visfatin in fat tissues of lean compared with fat control pigs. Domestic Animals Endrocrinology 35:58-73.
18. Baldassarre H, Hockley DK, Olaniyan B, Brochu E, Zhao X, Mustafa A and Bordignon V, 2008. Milk composition studies in transgenic goats expressing recombinant human butyrylcholinesterase in the mammary gland. Transgenic research 17:863-72.
19. Chappaz E, Albornoz MS, Campos D, Che L, Palin MF, Murphy BD, Bordignon V, 2008. Adiponectin enhances in vitro oocyte maturation and early embryo development in pigs. Domestic Animals Endrocrinology 35:198-207.
20. Ferreira R, Coelho de Oliveira JF, Antoniazzi AQ, Pimentel CA, Ferrugem Moraes JC, Henkes LH, Bordignon V, Gonçalves PBD, 2008. Relationship between clinical and postmortem evaluation in repeat breeder beef cows. Ciência Rural 38:1056-1060.

In: Proteins Researcher Biographical Sketches … ISBN: 978-1-62100-777-7
Editors: H. Z. Wang and M. Tian

Chapter 8

BIOGRAPHICAL SKETCH

NAME:	TITLE:	DATE OF BIRTH:
Tak Mao Chan	**PROFESSOR** University of Hong Kong, Hong Kong	**AUGUST 14, 1961**

EDUCATION:

Institution and Location	Degree	Year Conferred	Scientific Field
University of Hong Kong	MBBS	1985	

CONTACT POINTS:

Address: Department of Medicine, Room 303 New Clinical Building, Queen Mary Hospital, Pokfulam, Hong Kong

Professional Appointments:

1986-1991: Medical Officer, Medical & Health Department / Hospital Authority, Queen Mary Hospital / Tung Wah Hospital
1990-1991: Research Fellow, Renal Unit, Guy's Hospital, London, U.K.
1991-1993: Senior Medical Officer, Queen Mary Hospital
1994-1996; Lecturer, Department of Medicine, University of Hong Kong, Queen Mary Hospital
1997-1999: Associate Professor
1999-2004; Professor (Reader)
Since 2005: Professor (Personal Chair)

Since 2008 Yu Professor in Nephrology (Endowed Professorship)
Associate Dean (External Affairs), Faculty of Medicine
Since 2010 Chief of Clinical Service, Department of Medicine, Queen Mary Hospital and Hong Kong West Cluster Hospitals

Professional Experience, Honors and Key Offices

1) FRCP(London), FRCP(Edinburgh), FRCP(Glasgow), FHKCP, FHKAM, FASN.
2) Research awards - Patrick Manson Gold Medal for MD thesis, 1995; Outstanding Young Researcher Award, University of Hong Kong, 1999; Outstanding Researcher Award, University of Hong Kong, 2006.
3) President, Asian Pacific Society of Nephrology, 2008-2010; Chairman of Nephrology Board, Hong Kong College of Physicians.

Research Experience:

1. Research themes

i. Treatment and pathogenetic mechanisms of lupus nephritis. Key publications include:

Chan TM, et al. Identification of endothelial cell membrane proteins that bind human anti-DNA antibodies. *J Autoimmunity* 1997;10:433-9.
Chan TM, et al. Efficacy of mycophenolate mofetil in diffuse proliferative lupus nephritis. *N Engl J Med* 2000;343:1156-62.
Chan TM, Leung JKH, Ho SKN, Yung S. Mesangial cell-binding anti-DNA antibodies in patients with systemic lupus erythematosus. *J Am Soc Nephrol* 2002;13:1219-29.
Chan TM, et al. Long-term outcome of patients with diffuse proliferative lupus nephritis treated with prednisolone and oral cyclophosphamide followed by azathioprine. *Lupus* 2005;14:265-72.
Chan TM, et al. Long-term study of mycophenolate mofetil as both induction and maintenance treatment for diffuse proliferative lupus nephritis. *J Am Soc Nephrol* 2005;16:1076-84.

ii.Clinical course and management of viral hepatitis B / C in patients with renal failure. Key publications include:

Chan TM, et al. A prospective study of hepatitis C virus infection among renal transplant recipients. *Gastroenterology* 1993;104:862-8.
Chan TM, et al. Clinico-pathologic features of hepatitis C virus (HCV) infection in renal allograft recipients. *Transplantation* 1994; 58: 996-1000.
Chan TM, et al. Pre-emptive lamivudine therapy based on HBV DNA level in HBsAgpositive kidney allograft recipients. *Hepatology* 2002;36:1246-52.

Chan TM, et al. A prospective study on lamivudine-resistant hepatitis B in renal allograft recipients. *Am J Transplantation* 2004;4:1103-9. Chan TM. Hepatitis B and renal disease. *Curr Hepatitis Rep* 2010;9:99-105.

iii.Extracellular matrix and fibrosis. Key publications include:

Chan TM, et al, and Yung S. Emodin ameliorates glucose-induced matrix synthesis in human peritoneal mesothelial cells. *Kidney Int* 2003;64:519-33.

Yung S, et al. Reduction of perlecan synthesis and induction of TGF-beta1 in human peritoneal mesothelial cells due to high dialysate glucose concentration - implication in peritoneal dialysis. *J Am Soc Nephrol* 2004;15:1178-88.

Yung S, et al. Elevated glucose concentrations induce thrombospondin-1 synthesis in proximal renal tubular epithelial cells thereby stimulating fibronectin synthesis through TGFbeta1 dependent and TGF-beta1 independent pathways. *Nephrol Dial Transplant* 2006;21:1504-13.

2. Competitive grants and research postgraduate students

i.Track record of competitive earmarked research grants from the Research Grants Council of the Hong Kong Government.

ii.Track record of PhD and MPhil students supervised.

Publications during Last Three Years:

1) Yung S, Chan TM. Anti-DNA antibodies in the pathogenesis of lupus nephritis - the emerging mechanisms. *Autoimmun Rev* 2008; 7: 317-321.
2) Lui SL, Yung S, Tsang R, Zhang F, Chan KW, Tam S, Chan TM. Rapamycin prevents the development of nephritis in lupus-prone NZB/W F1 mice. *Lupus* 2008; 17: 305-313.
3) Lui SL, Yung S, Tsang R, Zhang F, Chan KW, Tam S, Chan TM. Rapamycin attenuates the severity of established nephritis in lupus-prone NZB/W F1 mice. *Nephrol Dial Transplant* 2008; 23: 2768-2776.
4) Chan TM. Mycophenolate mofetil in the treatment of lupus nephritis - 7 years on. *Lupus* 2008; 17: 617-621. [*Eminence-Based Medicine* Series]
5) Tse KC, Tang CSO, Lam MF, Yap DYH, Chan TM. Cost comparison between mycophenolate mofetil and cyclophosphamide-azathioprine in the treatment of lupus nephritis.*J Rheumatology* 2009; 36(1): 76-81.
6) Lui SL, Tsang R, Chan KW, Zhang F, Tam S, Yung S, Chan TM. Rapamycin attenuates the severity of murine adriamycin nephropathy. *Am J Nephrol* 2008; 29: 342-352.
7) Yung S, Chan TM. Intrinsic cells: mesothelial cells - central players in regulating inflammation and resolution. *Perit Dial Int* 2009; 29(S2): S21-27.
8) Yang W, Ng P, Zhao M, Hirankarn N, Lau CS, Mok CC, Chan TM, Wong RWS, Lee KW, Mok MY, Wong SN, Avihingsanon Y, Lee TL, Ho MHK, Lee PPW, Wong WHS, Lau YL. Population differences in SLE susceptibility genes: STAT4 and BLK, but not

PXK, are associated with systemic lupus erythematosus in Hong Kong Chinese. *Genes Immun* 2009; 10: 219-226.

9) Yung S, Zhang Q, Zhang J, Chan KW, Lui SL, Chan TM. Anti-DNA antibody induces PKC phosphorylation and fibronectin synthesis in human and murine lupus and the effect of mycophenolic acid. *Arthritis Rheum* 2009, 60: 2071-2082.
10) Yang W, Zhao M, Hirankarn N, Lau CS, Mok CC, Chan TM, Wong RWS, Lee KW, Mok MY, Wong SN, Avihingsanon Y, Lin IO, Lee TL, Ho MH, Lee PP, Wong WHS, Sham PC, Lau YL. ITGAM is associated with disease susceptibility and renal nephritis of systemic lupus erythematosus in Hong Kong Chinese and Thai. *Hum Mol Genet* 2009; 18: 2063-2070.
11) Yung S, Cheung KF, Zhang Q, Chan TM. Anti-DNA antibodies from patients with lupus nephritis bind to annexin II of human mesangial cells. *J Am Soc Nephrol* 2010; 21: 1912-1927.
12) Yang W, Shen N, Ye DQ, Liu Q, Zhang Y, Qian XX, Hirankarn N, Ying D, Pan HF, Mok CC, Chan TM, Wong RW, Lee KW, Mok MY, Wong SN, Leung AM, Li XP, Avihingsanon Y, Wong CM, Lee TL, Ho MH, Lee PP, Chang YK, Li PH, Li RJ, Zhang L, Wong WH, Ng IO, Lau CS, Sham PC, Lau YL; Asian Lupus Genetics Consortium. Genome-wide association study in Asian populations identifies variants in ETS1 and WDFY4 associated with systemic erythematosus. *PLoS Genet* 2010; 30: 187-191.
13) Chan TM. Mycophenolate mofetil as treatment for lupus nephritis. In: *Lupus Nephritis* Lewis EJ, Schwartz MM, Korbet SM and Chan DTM (eds) Oxford University Press, U.K., 2010, pp237-249.
14) Zhang Y, Yang W, Mok CC, Chan TM, Wong RW, Mok MY, Lee KW, Wong SN, Leung AM, Lee TL, Ho MH, Lee PP, Wong WH, Yang J, Zhang J, Wong CM, Ng IO, Garcia-Barceló MM, Cherny SS, Tam PK, Sham PC, Lau CS, Lau YL. Two missense variants in UHRF1BP1 are independently associated with systemic lupus erythematosus in Hong Kong Chinese. *Genes Immun* 2011; 12: 231-234.
15) Yang J, Yang W, Hirankarn N, Ye DQ, Zhang Y, Pan HF, Mok CC, Chan TM, Wong RW, Mok MY, Lee KW, Wong SN, Leung AM, Li XP, Avihingsanon Y, Rianthavorn P, Deekajorndej T, Suphapeetiporn K, Shotelersuk V, Baum L, Kwan P, Lee TL, Ho MH, Lee PP, Wong WH, Zeng S, Zhang J, Wong CM, Ng IO, Garcia-Barceló MM, Cherny SS, Tam PK, Sham PC, Lau CS, Lau YL. ELF1 is associated with systemic lupus erythematosus in Asian populations. *Hum Mol Genet* 2011; 20: 601-607.

ISBN: 978-1-62100-777-7

Chapter 9

BIOGRAPHICAL SKETCH

NAME:	TITLE:	DATE OF BIRTH:
Gopal Chandra	AICTE EMERITUS FELLOW, CENTRE FOR RURAL & CRYOGENIC TECHNOLOGIES, ADAVPUR UNIVERSITY , KOLKATA-700032, INDIA (FORMERLY: EMERITUS MEDICAL SCIENTIST OF INDIAN COUNCIL OF MEDICAL RESEARCH AND DIRECTOR- GRADE SCIENTIST, INDIAN INSTITUTE OF CHEMICAL BIOLOGY, KOLKATA - 700032, INDIA	JANUARY 01, 1942

EDUCATION:

Institution and Location	Degree	Year Conferred	Scientific Field
Dacca University	Matriculation	1958	
Calcutta University	Intermediate (Science)	1960	
	B. Sc. (Chemistry Hons.)	1962 8th	

(Continued)

Institution and Location	Degree	Year Conferred	Scientific Field
	M. Sc. (Biochemistry)	1964 3rd	
	Ph. D. (Biochemistry)	1970	

CONTACT POINTS:

Land Phone: +91-33-24217363 and +91-33-24146965
Miblie Phone : +91 93397 84545
E.Mail -majumdergc42@yahoo.co.in
and majumdergc42@gmail.com

Research Experience:

Major contributions were made in the following areas of biomedical research:

[i] Milk enzymes and physico-chemical properties of buffalo milk proteins.
[ii] Molecular Biology of hormone action for gene expression in mammary gland in *vitro*.
[iii] Reproductive biology of testis and epididymis.
[iv] Biochemical basis of the initiation / regulation of sperm motility using goat epididymal spermatozoa as a model.
[v] Membrane Biology: Structure and Function:
Analysis of sperm membrane lipids, phospholipid asymmetry and fluidity.
Synchronous modulation of sperm external surface lectin and its endogenous receptor.
Regulation of the phosphorylation status of a sperm outer surface phosphoprotein through a novel coupled enzyme system consisting of a protein kinase and phosphoprotein phosphatase.
[vi] Enzymology with special reference to protein kinases, phosphoprotein phosphatases and glycosidases.
[vii] Biotechnology & Proteomics:
Identification of novel motility – promoting and inhibiting proteins from different sources, which will be useful for rectification of some of the problems of human male infertility, cattle breeding, conservation of extinct species and also contraception.
Development of a novel computer-based spectrophotometric sperm â€œvertical velocityâ€• measuring instrumental system that will be useful in human infertility clinics, animal breeding centres and research laboratories, etc.
Development of a novel sperm cryopreservation method using chemically-defined medium rather than the conventional complex media containing egg-yolk, skim milk, etc.

Number of Research Publications: 123 including 11 review articles

Patents Granted:

A) In India

A process for the purification of a new motility-promoting protein from buffalo serum: A slaughter house waste by G.C. Majumder, M. Mandal and S. Banerjee.

Patent No 185383; Issue Date: August 3, 2001; Filing Date: March 17, 1997.

2) A process for the preparation of a purified new epididymal forward motility protein useful as a fertility promoter/blocker by G. C. Majumder and B. S. Jaiswal

Indian Patent No 189766; Issue Date: January 27, 2004; Filing Date: February 26, 1998.

B) In USA

1. Purified new epididymal forward motility protein and a process for the isolation of the said epididymal forward motility protein useful as a fertility promoter/blocker by G.C. Majumder and B.S. Jaiswal.

US Patent No. 6,231,862; Issue Date: May 15, 2001; Filing Date: January 12, 1998.

2. Purified new epididymal forward motility protein and a process for the isolation of the said epididymal forward motility protein useful as a fertility promoter/blocker by G.C. Majumder and B.S. Jaiswal

US Patent No 6306823; Issue Date: October 23, 2001; Filing Date: July 27, 1999.

3. A process for the purification of a new motility-promoting protein from buffalo serum: A slaughterhouse waste by G.C. Majumder, M. Mandal and S. Banerjee.

US Patent No 6613737; Issue Date: September 2, 2003; Filing Date: March 10, 1998.

C) In Japan

1) A process for the purification of a new motility-promoting protein from buffalo serum: A slaughterhouse waste by G.C. Majumder, M. Mandal and S. Banerjee.

Japan Patent No 3251545; Issue Date: November16, 2001; Filing Date: March 6, 1998.

2) Purified new epididymal forward motility protein and a process for the isolation of the said epididymal forward motility protein useful as a fertility promoter/blocker by G.C. Majumder and B.S. Jaiswal. Japan Patent No 3400367; Issue Date: Feb 21, 2003; Filing Date: Dec 24, 1998.

Professional Experience

Centre for Rural & Cryogenic Technologies,
Jadavpur University , Kolkata700032,
Vidyasgar University, Dept Human Physiology, Midnapur

AICTE Emeritus Fellow,
Guest Faculty - November,2007
March 2007 – Present
Present - Research work on sperm cryobiology

Teaching Reproductive Endocrinology to Postgraduate Students
Indian Institute of Chemical Biology (IICB), Kolkata
Emeritus Medical Scientist, (ICMR)
Nov 1 2002
December31, 2006

Biochemical basis of sperm motility initiation/regulation
University of Wyoming, USA
Visiting Scientist
July 2, 2002
Oct 28, 2002
Sperm Chemotaxis

Indian Institute of Chemical Biology (IICB), Kolkata
Scientist â€˜Gâ€™
(Director Grade)
April 1994
Retired on 31.12.2001

Biochemical basis of sperm motility initiation/regulation
IICB, Kolkata
Scientist â€˜Fâ€™
(Deputy Director)
April 1989
April 1994

IICB, Kolkata
Scientist â€˜E-IIâ€™
(Assistant Director)
April 1986
April 1989

IICB, Kolkata
Scientist â€˜E-Iâ€™
April 1981
April 1986

IICB, Kolkata
Scientist â€˜Câ€™
April 1975
April 1981

St. Luke's Hospital, Milwaukee, USA
Medical Investigator
July 1973
Mar. 1975

Reproductive biology of epididymis and mammary gland
University of Wisconsin, Madison, USA
Research Associate
June 1971
June 1973

Gene expression in testis and mammary gland
Duke University, North Carolina, USA
Post-doctoral Research Fellow
May 1970
May 1971

Hormonal regulation of gene expression in mammary gland
National Dairy Research Institute, Karnal, Harayana
Research Assistant & Sr. Tech. Assistant
Feb. 1965
May 1970

Buffalo milk proteins, milk enzymes *etc.*
School of Tropical Medicine, Kolkata
JRF/CSIR
Sept. 1964
Feb. 1965
Preliminary research work

Other Foreign Travels for Academic Accomplishments:

[i] **Japan:** Participated the first AMBO course on "Fluorescent probes for the analysis of cell structure" held in Kyoto University, Japan during the period October 12-24, 1981(Also mentioned above).

[ii] **USSR:** Visited several Research Institutes at Moscow and Leningrad of USSR under Indian National Science Academy and USSR - Academy of Science Exchange Programme during the period Sept. 10 - October 19, 1985. Institutes visited at Moscow were - Institute of Molecular Biology, Institute of Bio organic Chemistry, Institute of Cytology and Dept. of Biochemistry, Moscow University. At Leningrad I visited Institute of Cytology and Institute of Animal Breeding and Genetics.

[iii] **Germany:** Under CSIR-DAAD Exchange Programme of Scientists during September 13 to November 12, 1995 (two months), I visited the Institute of Reproductive Medicine, Munster and Institute of Hormone and Fertility Research, Hamburg, in Germany.

[iv] **USA:** At the invitation of Prof R. W. Atherton, I have visited the Department of Zoology and Biochemistry, University of Wyoming, Laramie, USA as a Visiting Scientist for four months during the period: July-October, 2002

Honors:

[i] **Rafi Ahmed Kidwai Award:** In January 1976, I was awarded prestigious Rafi Ahmed Kidwai Memorial Prize for the biennium 1972-73 by Indian Council of Agricultural Research, Govt. of India.

[ii] **Elected Fellow of Academies:**

[a] Fellow National Academy of Sciences since 1998.

[b] Fellow of West Bengal Academy of Science & Technology since 1994.

Publications during Last Three Years:

1. Ghoshal S., Sengupta T., Dungdung S.R., Majumder G.C., Sen P.C. (2008). Characterization of a low molecular mass stimulator protein of Mg^{2+}-independent Ca^{2+}-ATPase: effect on phosphorylation-dephosphorylation, calcium transport and sperm cell motility. *Biosci Rep.* 28, 61-71
2. Maiti A., Mishra K.P., Majumder G.C.. (2008) Role of the major ecto-phosphoprotein in sperm flagellar motility using a cell electroporation method. *Mol Reprod Dev.* 75, 1185 – 1195
3. Sengupta T., Ghoshal S., Dungdung S.R., Majumder G.C., Sen P.C.. (2008) Structural and functional characterization and physiological significance of a stimulator protein of Mg^{2+} -independent Ca (2+)-ATPase isolated from goat spermatozoa. *Mol Cell Biochem. 3*11, 93-103
4. Nath D., Maiti A., Majumder G.C.. (2008). Cell surface phosphorylation by a novel ecto-protein kinase: a key regulator of cellular functions in spermatozoa. *Biochim Biophys Acta.* 1778:153-65
5. Maiti, A., Nath, D., Dungdung, S. R and Majumder, G. C (2009) Sperm ecto-protein kinase and its protein substrate: novel regulators of membrane fusion during acrosome reaction. *J. Cell. Physiol.* 220, 394-400 (2009).
6. Jaiswal B. S., Das K., Saha S., Dungdung S.R. and Majumder G.C. (2010) Purification and characterization of a motility initiating protein from caprine epididymal plasma.: *J. Cell. Physiol.*222:254-263
7. Banerjee S. and Majumder G. C. (2010).Homologous liver parenchymal cell-cell adhesion mediated by an endogenous lectin and its receptor. Cellu Mol. Biol. Letters. 15 , 356-364
8. Das, S. Saha, S., Majumder, G. C., Dungdung, S. R. (2010)Purification and characterization of a sperm motility inhibiting factor from caprine epididymal plasma *Plos One* 5, e12039, page 1-12
9. Kundu, C. N., Majumder, G. C, and Preet, R.(2010) Cryopreservation of Spermatozoa: Recent Biotechnological Advancement in Gamete Preservation Technology . In â€œBiomedical Engineering and Information Systems: Technologies, Tools and

Applicationsâ€• (Ed . A. Shukla and R. Tiwari), IGI -Global Publisher, Gwalior , page 277-302

10. Majumder G. C., Saha, S., Das, K, Nath, D., Maiti, A., Chakrabarty, J., Das, S., Bhoumik, A., Das, A., Ghosh P., Banerjee S., Mandal M., Jaiswal, B. S. Bhattacharyya, D and Dungdung S. R.: Biochemical role of cell surface molecules in the regulation of sperm flagellar movement. In *Endocrinology and Male Reproductive Biology (*Ed. S. Singh), *PHI* Learning, New Delhi (in press)

In: Proteins Researcher Biographical Sketches … ISBN: 978-1-62100-777-7
Editors: H. Z. Wang and M. Tian

Chapter 10

BIOGRAPHICAL SKETCH

NAME:	TITLE:	DATE OF BIRTH:
Laura Díez González	**DEPARTAMENT OF OTOLARINGOLOGY, BIERZO HOSPITAL, LEON, ESPAÑA.**	**4/4/1980**

EDUCATION:

Institution and Location	Degree	Year Conferred	Scientific Field
University of Oviedo	Degree in medicine and surgery	1998-2004	Otolaryngologist.

CONTACT POINTS:

Address: C/ La Paz, 6, 4B, Ponferrada (León).

RESEARCH AND PROFESSIONAL EXPERIENCE:

Diploma of Advanced Studies (D.E.A.), "Ciencias de la Salud" Programme, in the Genetics, Biochemistry and Immunology Department of the University of Vigo (2005-2007). *"Estudio de la eficacia de la detección precoz de la hipoacusia en el periodo neonatal en el Complexo Hospitalario Universitario de Vigo".*

Professional Appointments:

- Department of Otolaryngology at CHUVI Hospital.

- Department of Otorlaryngology at Bierzo Hospital.

Honors:

- Best General Scientific Session. Detección precoz de la hipoacusia neonatal. Comisión de docencia del Complexo Hospitalario Universitario de Vigo. Vigo 2008.
- Third Award. Clinical Case Competition 2009. Deformidad craneofacial. Sociedad Española de otorrinolaringología y patología cérvico facial.

Publications during Last Three Years:

- L. Díez, A. G. Prado; D. Sánchez. Otomastoiditis tuberculosa. A propósito de un caso.Revista portuguesa de otorrinolaringología y patología cérvicofacial. 2008 Sep; 43(3):203- 206.
- L. Díez, A. G. Prado; N. Acevedo. Traumatismos penetrantes laringotraqueales. Revista portuguesa de otorrinolaringología y patología cérvicofacial. 2008 Dic; 46(4):261- 264.
- L. Díez, A. Prado, D. Sánchez. Programa de detección precoz de la sordera en el periodo neonatal en el hospital xeral- cíes de vigo. Nuestros resultados. Revista Eído. 2009 vol.2.
- L. Díez, A. Quintana, O. Castro. Programa de detección precoz de la sordera neonatal.Revista portuguesa de otorrinolaringología y patología cérvicofacial. 2010 Sep;48(3): 131-134
- L. Díez, I. Guijarro, P. Vaamonde. Manifestación primaria de linfoma Hodgkin en adenoides. A propósito de un caso. Acta Otorrinolaringol Esp. 2010;61(6):462-464

In: Proteins Researcher Biographical Sketches … ISBN: 978-1-62100-777-7
Editors: H. Z. Wang and M. Tian

Chapter 11

BIOGRAPHICAL SKETCH

NAME:	TITLE:	DATE OF BIRTH:
Florence Edwards-Lévy	**INSTITUTE OF MOLECULAR CHEMISTRY OF REIMS, FACULTY OF PHARMACY OF REIMS, UNIVERSITY OF REIMS CHAMPAGNE-ARDENNE**	**SEPTEMBER 18TH 1966**

EDUCATION:

Institution and Location	Degree	Year Conferred	Scientific Field
	Pharm. D.	1991	
	Master in Chemistry	1991	
	Ph.D. in Pharmaceutical Technology	1994	

CONTACT POINTS:

Address:
Institut de Chimie Moléculaire de Reims, CNRS UMR6229
Faculté de Pharmacie
51 rue Cognacq-Jay
51100 Reims
France

RESEARCH AND PROFESSIONAL EXPERIENCE:

Associate Professor in Pharmaceutical Technology at the Faculty of Pharmacy of Reims since 1995.

Professional Appointments:

Associate Professor in Pharmaceutical Technology at the Faculty of Pharmacy of Reims since 1995.

Publications during Last Three Years:

M. Callewaert, D. Laurent-Maquin and F. Edwards-Lévy, Albumin-alginate coated microspheres: resistance to steam sterilization and to lyophilization. Int. J. Pharm. 2007, 344, 161.

F. Edwards-Lévy, Méthodes d'encapsulation par transacylation : exemples d'applications. In « Microencapsulation », Coordination T. Vandamme, D. Poncelet et P. Subra-Paternault, Lavoisier Tec&Doc, Paris, (2007), pp. 313-322.

Y. Lefebvre, E. Leclerc, D. Barthès-Biesel, J. Walter and F. Edwards-Lévy, Flow of artificial microcapsules in microfluidic channels: a method for determining the elastic properties of the membrane. Phys. Fluids. 2008, 20, 123102.

F. Edwards-Lévy, Bioencapsulation. In « Pharmacie Galénique, formulation et technologie pharmaceutique », Coordination P. Wehrlé, Maloine, Paris, (2008), pp.253-259.

M. Callewaert, J.-M. Millot, J. Lesage, D. Laurent-Maquin and F. Edwards-Lévy, Albumin-alginate microspheres: role of structure in binding and release of the KRFK peptide. Int. J. Pharm. 2009, 366 (1-2), 103-110.

Thi-Xuan Chu, Anne-Virginie Salsac, Eric Leclerc, Dominique Barthès-Biesel, Hélène Wurtz, Florence Edwards-Lévy, Comparison between measurements of elasticity and free amino group content of ovalbumin microcapsule membranes: discrimination of the cross-linking degree. J. Coll. Interf. Sci. 2011, 355 (1), 81-88.

In: Proteins Researcher Biographical Sketches … ISBN: 978-1-62100-777-7
Editors: H. Z. Wang and M. Tian

Chapter 12

BIOGRAPHICAL SKETCH

NAME:	TITLE:	DATE OF BIRTH:
Sergei Fedorovich	DR. INSTITUTE OF BIOPHYSICS AND CELL ENGINEERING	28 AUGUST 1969

EDUCATION:

Institution and Location	Degree	Year Conferred	Scientific Field
	PhD		Biophysics

CONTACT POINTS:

Address: Akademicheskaya St., 27, Minsk, Belarus

Professional Appointments:

Group leader, Institute of Biophysics and Cell Engineering, Minsk, Belarus

Honors:

2010- Matsumae fellowship (Japan)
2009- Landau-Volta Network Fellowship (Italy)
2008- DAAD fellowship (Germany)
2006- The Physiological Society, "Centre of excellence" award.
2005- EMBO short-term fellowship

2003- Wellcome Trust International Research Development Award
1999- Wellcome Trust Travelling Research Fellowship
1998- 3rd degree Diploma of best post-graduate student's articles competition
1997- Special Bursary of Belorussian Government for post-graduate students
1997- 2nd degree Diploma of best post-graduate student's articles competition

Publications during Last Three Years:

1. Waseem T.V., Fedorovich S.V. (2010) Presynaptic glycine receptors influence plasma membrane potential and glutamate release. Neurochem. Res., 35, 1188-1195.
2. Fedorovich S.V., Alekseenko A.V., Waseem T.V. (2010) Are synapses targets of nanoparticles? Biochem. Soc. Trans., 38, 536-538.
3. Alekseenko A.V., Waseem T.V., Fedorovich S.V. (2008) Ferritin, a protein containing iron nanoparticles, induces reactive oxygen species formation and inhibits glutamate uptake in rat brain synaptosomes. Brain Res., 1241, 193-200
4. Waseem T.V., Lapatsina L.P., Fedorovich S.V. (2008) Influence of integrin-blocking peptide on gadolinium and hypertonic shrinking-induced neurotransmitter release in rat brain synaptosomes. Neurochem. Res. 33, 1316-1324.

In: Proteins Researcher Biographical Sketches ... ISBN: 978-1-62100-777-7
Editors: H. Z. Wang and M. Tian

Chapter 13

BIOGRAPHICAL SKETCH

NAME:	TITLE:	DATE OF BIRTH:
Celeste C. Finnerty	**PHD SHRINERS HOSPITALS FOR CHILDREN, INSTITUTE FOR TRANSLATIONAL SCIENCES AND SEALY CENTER FOR MOLECULAR MEDICINE, AND DEPARTMENT OF SURGERY, UNIVERSITY OF TEXAS MEDICAL BRANCH, GALVESTON, TEXAS, U.S.A.**	**4/29/71**

CONTACT POINTS:

Address: 815 Market St, Shriners Hospitals for Children, Galveston, TX 77550

RESEARCH AND PROFESSIONAL EXPERIENCE:

2001 – 2003 Fellowship, Sealy Center for Cancer Cell Biology, University of Texas Medical Branch, Galveston, TX

2003 – 2006 NIH Postdoctoral Research Fellow in Trauma and Burns, Shriners Hospital for Children / Department of Surgery, University of Texas Medical Branch, Galveston, TX

2005 – 2007 Instructor, Department of Surgery, University of Texas Medical Branch, Galveston, TX

2006 – 2009 Shrine Research Fellowship, Shriners Hospital for Children / Department of Surgery, University of Texas Medical Branch, Galveston, TX
2007 – present Assistant Professor, Department of Surgery, University of Texas Medical Branch, Galveston, TX
2008 – present Scholar, Clinical Research Scholars Program, Clinical Research Education Program, University of Texas Medical Branch, Galveston, TX
2008 – 2010 Assistant Coordinator of Research, Department of Surgery Burn Unit, Galveston, TX
2008 – present Cell Biology Graduate Program Faculty, University of Texas Medical Branch, Galveston, TX
2010 – presentt Sealy Center for Molecular Medicine Faculty, University of Texas Medical Branch, Galveston, TX
2010 – present Associate Director of Research, Department of Surgery Burn Unit / Shriners Hospitals for Children, Galveston, TX

Professional Appointments:

Professional Appointments: Assistant Professor in Surgery, Associate Director of Research

Honors:

2010 3M Wound Healing Society Foundation Fellowship Recipient
2008-present Clinical Research Scholar, Clinical Scholar Research program, UTMB
2007 Research Citation Finalist, 36th Critical Care Congress
2006 -- 2008 Shrine Research Fellow
2004 – 2006 NIH Postdoctoral Research Fellow in Traumma and Burns
2005 1st Place Poster Presentation, Biochemistry and Molecular Biology, UTMB
2003 UTMB Post-doctoral Representative for Capitol Hill Day
2000 Katherina Siebert Award for Excellence in Oncologic Research
2000 Elias Hochman Research Award for Advances in Toxicologic Research
1999 1st Place Platform Presentation, Gulf Coast Society of Toxicology Annual Meeting
1998 Arthur V. Simmang Academic Scholarship.
1998 Centennial Center of Environmental Toxicology Award
1997 – 2000 NIEHS Toxicology Pre-doctoral Fellowship
1996 Zelda Zinn Casper Award for Research in the Biomedical Sciences

Publications during Last Three Years:

1. Gauglitz GG, Song J, Herndon DN, Finnerty CC, Boehning DF, Barral JM, Jeschke MG. Characterization of the inflammatory response during acute and post-acute phases after severe burn. Shock. 2008 Nov: 30(5) 503-7.

2. Jeschke MG, Norbury WB, Finnerty CC, Mlcak RP, Kulp GA, Branski LK, Gauglitz GG, Herndon B, Swick A, Herndon DN. Age differences in inflammatory and hypermetabolic postburn responses. Pediatrics. 2008 Mar, 121(3) 497-507.
3. Jeschke MG, Finnerty CC, Kulp GA, Przkora R, Mlcak RP, Herndon DN. Combination of recombinant human growth hormone and propranolol decreases hypermetabolism and inflammation in severely burned children. Pediatric Critical Care Medicine. 9(2):209-216, March 2008.
4. Russom A, Sethu P, Irimia D, Mindrinos MN, Calvano SE, Garcia I, Finnerty C, Tannahill C, Abouhamze A, Wilhelmy J, LÃ³pez MC, Baker HV, Herndon DN, Lowry SF, Maier RV, Davis RW, Moldawer LL, Tompkins RG, Toner M; the Inflammation and Host Response to Injury Large Scale Collaborative Research Program.Microfluidic Leukocyte Isolation for Gene Expression Analysis in Critically Ill Hospitalized Patients. Clin Chem. 2008 May;54(5):891-900.
5. Jeschke MG, Chinkes DL, Finnerty CC, Kulp G, Suman OE, Norbury WB, Branski LK, Gauglitz GG, Mlcak RP, Herndon DN. Pathophysiologic response to severe burn injury. Annals of Surgery. 248(3):387-401, 2008.
6. Gauglitz GG*, Finnerty CC*, Herndon DN, Mlcak RP, Jeschke MG. Are serum cytokines early predictors for the outcome of burn patients with inhalation injuries who do not survive? Critical Care. 12(3): R81 (Epub ahead of print), June 2008. * - shared first authorship
7. Jeschke MG, Mlcak RP, Finnerty CC, Norbury WB, Przkora R, Kulp GA, Gauglitz GG, Zhang XJ, Herndon DN. Gender differences in pediatric burn patients: Does it make a difference? Annals of Surgery, July 2008. 248(1):126-36.
8. Finnerty CC, Jeschke MG, Herndon DN, Gamelli RL, Gibran NS, Klein MB, Silver GM, Arnoldo B, Remick DG, Tompkins RG, and The Inflammation and the Host Response to Injury Collaborative Research Program. Temporal cytokine profiles in severely burned patients: a comparison of adults and children. Molecular Medicine Sept-Oct 2008; 14(9-10) 553-560.
9. Finnerty CC, Przkora R, Herndon DN, Jeschke MG. Cytokine expression profile over time in burned mice. Cytokine. 45(1):20-5, 2009 Jan.
10. Jeschke MG, Gauglitz GG, Song J, Kulp GA, Finnerty CC, Cox RA, Barral JM, Herndon DN, Boehning D. Calcium and ER stress mediate hepatic apoptosis after burn injury. J Cell Mol Med Aug 2009; 13(8B):1857-65.
11. Song J, Wolf SE, Wu XW, Finnerty CC, Gauglitz GG, Herndon DN, Jeschke MG. Starvation-induced proximal gut mucosal atrophy diminished with aging. Journal of Parenteral and Enteral Nutrition. 2009 Jul-Aug;33(4):411-6. Epub 2009 Jan 6.
12. West MA, Moore EE, Shapiro MB, NAthens AB, Cuschieri J, Johnson JL, Harbrecht BG, Minei JP, Bankey PE, Maier RV, Inflammation and the Host Response to Injury Collaborative Research Program. Inflammation and the host response to injury, a large-scale collaborative project: patient-priented research core—standard operatiing procedures for clinical care VII—Guidelines for anttibiotic administration in severely injured patients. J Trauma. 2008 Dec; 65(6):1511-9.
13. Qian WJ, Liu T, Petyuk V, Gritsenko M, Polpitiya A, Kaushal A, Xiao W, Finnerty C, Jeschke M, Monroe M, Moore R, Moldawer L, Davis R, Tompkins R, Herndon D, Camp D, Smith R, Petritis B. Large-Scale Multiplexed Quantitative Discovery Proteomics

Enabled by the Use of an 18O-labeled Universal Reference Sample. Journal of Proteome Research. 2009 Jan; 8(1):290-9.

14. Pham TN, Klein MB, Gibran NS, Arnoldo BD, Gamelli RL, Silver GM, Jeschke MG, Finnerty CC, Tompkins RG, Herndon DN. Impact of oxandrolone treatment on acute outcomes after severe burn injury. J Burn Care Res. 2008 Nov-Dec;29(6):902-6.
15. Zhang XJ, Meng C, Chinkes DL, Finnerty CC, Aarsland AA, Jeschke MG, Herndon DN. Acute propranolol infusion stimulates protein synthesis in skin wound. Surgery. 2009 May; 145(5):558-67. Epub 2009 Mar 21.
16. Song J, Finnerty CC, Herndon DN, Boehning D, Jeschke MG. Severe Burn-Induced Endoplasmic Reticulum Stress and Hepatic Damage in Mice. Mol Med. 2009 Jul 15. [Epub ahead of print].
17. Warren HS, Elson CM, Hayden DL, Schoenfeld DA, Cobb JP, Maier RV, Moldawer LL, Moore EE, Harbrecht BG, Pelak K, Cuschieri J, Herndon DN, Jeschke MG, Finnerty CC, Brownstein BH, Hennessy L, Mason PH, Tompkins RG; Inflammation and Host Response to Injury Large Scale Collaborative Research Program. A genomic score prognostic of outcome in trauma patients. Mol Med. 2009 Jul-Aug;15(7-8):220-7. Epub 2009 Apr 10.
18. Hayden D, Lazar P, Schoenfeld D; Inflammation and the Host Response to Injury Investigators. Assessing statistical significance in microarray experiments using the distance between microarrays. PLoS One. 2009 Jun 16;4(6):e5838.
19. Rajicic N, Finkelstein DM, Schoenfeld DA; Inflammation and Host Response to Injury Research Program Investigators. Analysis of the relationship between longitudinal gene expressions and ordered categorical event data. Stat Med. 2009 Sep 30;28(22):2817-32.
20. Williams FN, Herndon DN, Hawkins HK, Lee JO, Cox RA, Kulp GA, Finnerty CC, Chinkes DL, Jeschke MG. The leading causes of death after burn injury in a single pediatric burn center. Crit Care. 2009 Nov 17; 13(6):R183.
21. Zhou B. Xu W. Herndon D. Tompkins R. Davis R. Xiao W. Wong WH. Inflammation and Host Response to Injury Program. Analysis of factorial time-course microarrays with application to a clinical study of burn injury. Proceedings of the National Academy of Sciences of the United States of America. 2010 Jun 1: 107(22):9923-8.
22. Kotz KT, Xiao W, Miller-Graziano C, Qian WJ, Russom A, Warner EA, Moldawer LL, De A, Bankey PE, Petritis BO, Camp DG 2nd, Rosenbach AE, Goverman J, Fagan SP, Brownstein BH, Irimia D, Xu W, Wilhelmy J, Mindrinos MN, Smith RD, Davis RW, Tompkins RG, Toner M; Inflammation and the Host Response to Injury Collaborative Research Program. Clinical microfluidics for neutrophil genomics and proteomics. Nat Med. 2010 Sep;16(9):1042-7. Epub 2010 Aug 29.PMID: 20802500
23. Qian WJ*, Petritis BO*, Kaushal A*, Finnerty CC*, Jeschke MG*, Monroe ME, Moore RJ, Schepmoes AA, Xiao W, Moldawer LL, Davis RW, Tompkins RG, Herndon DN, Camp DG 2nd, Smith RD; Inflammation and the Host Response to Injury Large Scale Collaborative Research Program. Plasma proteome response to severe burn injury revealed by 18O-labeled "universal" reference-based quantitative proteomics. J Proteome Res. 2010 Sep 3;9(9):4779-89.PMID: 20698492 * - shared first authorship
24. Song J, Wolf SE, Wu XW, Finnerty CC, Herndon DN, Jeschke MG. Proximal Gut Mucosal Epithelial Homeostasis in Aged IL-1 Type I Receptor Knockout Mice after Starvation. J Surg Res. 2010 Apr 21. [Epub ahead of print]PMID: 20605606

25. Jeschke MG, Kulp GA, Kraft R, Finnerty CC, Mlcak R, Lee JO, Herndon DN. Intensive insulin therapy in severely burned pediatric patients: a prospective randomized trial. Am J Respir Crit Care Med. 2010 Aug 1;182(3):351-9. Epub 2010 Apr 15.PMID: 20395554
26. Traber MG, Leonard SW, Traber DL, Traber LD, Gallagher J, Bobe G, Jeschke MG, Finnerty CC, Herndon DN. Î±-Tocopherol adipose tissue stores are depleted after burn injury in pediatric patients. Am J Clin Nutr. 2010; 92(6):1378-1384. PMID 20881067
27. Rajicic N, Cuschieri J, Finkelstein DM, Miller-Graziano CL, Hayden D, Moldawer LL, Moore E, O'Keefe G, Pelik K, Warren HS, Schoenfeld DA; Inflammation and the Host Response to Injury Large Scale Collaborative Research Program. Identification and interpretation of longitudinal gene expression changes in trauma. PLoS One. 2010 Dec 20;5(12):e14380.PMID:21187951

In: Proteins Researcher Biographical Sketches …
Editors: H. Z. Wang and M. Tian
ISBN: 978-1-62100-777-7

Chapter 14

BIOGRAPHICAL SKETCH

NAME:	TITLE:	DATE OF BIRTH:
Yuchang Fu	ASSOCIATE PROFESSOR	

EDUCATION:

Institution and Location	Degree	Year Conferred	Scientific Field
Shanghai University, Shanghai, China	BS	1982	Biology
Chinese Academy of Science, Wuhan, China	MS	1985	Genetics
Kyushu University, Fukuoka, Japan	PhD	1990	Molecular Biology

A. Personal Statement

The goal of the proposed research is to study the mechanisms of AdipoR1 gene expression and regulation to improve atherogenesis and metabolic syndrome in animal models. Specifically, we plan to investigate the functions of AdipoR1 gene to lipid metabolism and inflammatory response in macrophages related to atherosclerosis and metabolic syndrome.

I have the expertise and motivation necessary to successfully carry out the proposed work based on the fact that I have a broad background in molecular and cell biology, with specific training and expertise in key research areas for this application. I have been involved in research projects that investigate molecular mechanisms of LDL receptor, lipid binding protein, NR4A nuclear receptors and adiponectin gene function and regulation in cell and animal models. I have successfully supervised these research projects and produced several peer-reviewed publications from each project. The current application is very highly related to my prior work, and this study will provide novel therapeutic applications and treatments for preventing atherosclerosis and metabolic syndrome. In summary, I have a demonstrated record of successful and productive research projects in an area of high relevance for our society, and my expertise and experience have prepared me to effectively perform the proposed project.

B. Positions and Honors

Positions and Employment

1990-1993 Research Associate, JSR Research Institute, Tsukuba, Japan
1993-1994 Research Fellow, Beth Israel Hospital, Harvard Medical School, Boston, MA
1995-1998 Research Fellow, University of California, San Diego, CA
1998-2003 Research Assistant Professor, Medical University of South Carolina, Charleston, SC
2003-2010 Assistant Professor (tenure-track), University of Alabama at Birmingham, Birmingham, AL
2006- Senior Scientist, UAB Diabetes Research and Training Center (DRTC)
2010- Scientist, UAB Center for Cardiovascular Biology (CCVB)
2010- Associate Professor, University of Alabama at Birmingham, Birmingham, AL

Other Experience and Professional Memberships

1986-1990 Member, Japanese Molecular Biology Association
1986-1990 Member, Japanese Genetics Association
1986-1993 Member, Japanese Development Biology Association
1993- Member, American Heart Association
2003- Member, American Diabetes Association
2008- Ad Hoc Reviewer of American Diabetes Association
2009- Member, International Atherosclerotic Society
2009- Member, Research Grant Review Committee, American Diabetes Association
2011- Member, Research Grant Review Committee, American Heart Association

Honors and Awards

1979 Excellent Student of the Year
1981 Excellent Student of the Year
1982-1985 Fellowship Award, Chinese Academy of Science
1986-1987 Scholarship Award, The Ministry of Education, Science and Culture of Japan
1988-1989 Scholarship Award, The Ministry of Education, Science and Culture of Japan

C. Selected Peer-reviewed Publications (Selected from 32 peer-reviewed publications)

Most relevant to the current application (in chronological order)

1. *Fu, Y., Luo, N., Klein, R. L., and Garvey, W. T. (2005). Adiponectin promotes adipocyte differentiation, Insulin sensitivity, and lipid accumulation. *J. Lipid Research.* 46, 1369-1379. PMID: 15834118
2. *Fu, Y., Luo, L., Luo, N., and Garvey, W. T. (2006). Lipid metabolism mediated by adipocyte lipid binding protein (ALBP/aP2) gene expression in human THP-1 macrophages. *Atherosclerosis.* 188, 102-111. PMID: 16313911
3. *Fu, Y., Luo, L., Luo, N., and Garvey, W. T. (2006). Proinflammatory cytokine production and insulin sensitivity regulated by overexpression of resistin in 3T3-L1 adipocytes. *Nutrition & Metabolism.* 3, 28. PMID: 16854242
4. *Fu, Y., Luo, L., Luo, N., Zhu, X and Garvey, W. T. (2007). NR4A orphan nuclear receptors modulate insulin action and the glucose transport system: potential role in insulin resistance. *J. Biol. Chem.* 282, 31525-31533. PMID: 17785466
5. Tian, L., Luo, N., Klein, R. L., Chung, B. H., Garvey, W. T., and Fu, Y. (2009). Adiponectin reduces lipid accumulation in macrophage foam cells. *Atherosclerosis.* 202, 152-161. PMID: 18511057
6. Luo, N., Liu, J., Chung, B. H., Yang, Q-L., Klein, R. L., Garvey, W. T., and Fu, Y. (2010). Macrophage adiponectin expression improves insulin sensitivity and protects against inflammation and atherosclerosis. *Diabetes.* 59, 791-799. PMID: 20350970
7. Yin, K., Deng, X., Mo, Z-C., Zhao, G-J., Jiang, J., Cui, L-B., Tan, C-Z., Wen, G-B., Fu, Y., and Tang, C-K. (2011). Tristetraprolin-dependent posttranscriptional regulation of inflammatory cytokines mRNA expression by apolipoprotein A-I: role of ATP-binding membrane cassette transporter A1 and signal transducer and activator of transcription 3. *J. Biol. Chem.* 286, 13834-13845.
8. Luo, N., Wang, X., Chung, B. H., Lee, M-H., Klein, R. L., Garvey, W. T., and Fu, Y. (2011). Effects of macrophage-specific adiponectin expression on lipid metabolism *in vivo. AJP-Endocrinology and Metabolism.* In press.

Additional recent publications of importance to the field (in chronological order)

1. Fu, Y., Wang, Y., and Evans, S. M. (1998). Viral sequences enable efficient and tissue specific expression of transgenes in *Xenopus. Nature Biotechnology.* 16, 253-257. PMID: 9528004
2. Fu, Y., Yan, W., Mohun, T., and Evans, S. M. (1998). Vertebrate *tinman* homologues *XNkx2-3* and *XNkx2-5* are required for heart formation in a functionally redundant manner. *Development.* 125, 4439-4449. PMID: 9778503
3. *Fu, Y., Luo, N., and Lopes-Virella, M. F. (2000). Oxidized LDL induces the expression of adipocyte lipid binding protein (ALBP/aP2) mRNA and protein in human THP-1 macrophages. *J. Lipid Research.* 41, 2017-2023. PMID: 11108735
4. *Fu, Y., Luo, N., and Lopes-Virella, M. F. (2002). Upregulation of interleukin-8 expression by prostaglandin D2 metabolite 15-deoxy-delta12, 14 prostaglandin J2 (15d-PGJ2) in human THP-1 macrophages. *Atherosclerosis.* 160, 11-20. PMID: 11755918

5. *Fu, Y., Luo, N., Lopes-Virella, M. F., and Garvey, W. T. (2002). The adipocyte lipid binding protein (ALBP/aP2) gene facilitates foam cell formation in human THP-1 macrophages. *Atherosclerosis.* 165, 259-269. PMID: 12417276
6. Fu, Y., Huang, Y., Bandyopadhyay, S., Virella, G., and Lopes-Virella, M. F. (2003). LDL immune complexes stimulate low density lipoprotein receptor expression in U937 histiocytes via extracellular signal-regulated kinase and AP-1. *J. Lipid Research.* 44, 1315-1321. PMID: 12730303
7. *Fu, Y., Maianu, L., Melbert, B. R., and Garvey, W. T. (2004). Facilitative Glucose Transporter Gene Expression in Human Lymphocytes, Monocytes, and Macrophages: a Role for GLUT Isoforms 1, 3, and 5 in the Immune Response and Foam Cell Formation. *Blood Cells, Molecules, & Diseases.* 32, 182-190. PMID: 14757434

*** As the first and corresponding author**

D. Research Support

<u>Ongoing Research Support</u>

Research Award, Fu and Garvey (PIs) 04/01/09-12/31/11
Alabama Drug Discovery Alliance (ADDA)
"NR4A3 orphan nuclear receptor as a target for the treatment of insulin resistance"
This proposal explores and develops strategies to identify small-molecule agonists for NR4A3 which is an orphan nuclear receptor expressed in human skeletal muscle. These small molecules are strong therapeutic candidates to treat human diseases caused by insulin resistance.
Role on Project: Co-Principal Investigator

R01 DK38765, Garvey (PI) 09/01/06-08/31/11
NIH/NIDDK
"Mechanisms of Human Insulin Resistance"
This proposal examines mechanisms and interrelationships between molecular defects in glucose and fat metabolism in skeletal muscle from insulin resistant humans, particularly as relates to mitochondrial function.
Role on Project: Co-Investigator

R01 DK083562, Garvey (PI) 07/01/09-06/30/14
NIH/NIDDK
"NR4A orphan receptors and insulin resistance"
This proposal examines the role of NR4A family of orphan nuclear receptors in modulating insulin action, their role in human insulin resistance, and their potential as a therapeutic drug target using a translational approach that includes studies in humans, transgenic mice, and cultured cell systems.
Role on Project: Co-Investigator

Research Award, Garber (PI) 10/01/09-09/31/11
UAB Diabetes Research and Training Center
"Effects of apolipoprotein-mimetic peptides on adipocyte/macrophage crosstalk"
This project investigates processes which lead to inflammation of adipose (fat) tissue, and the role of lipoproteins in modifying those processes, using small peptide molecules which mimic the structures of several of the major proteins present in lipoproteins.
Role on Project: Co-Investigator

Pending Research Support

GSA Spring 2011 Grant-in-Aid, Fu (PI) 07/01/11-06/30/13
AHA
"Reversal of Cardiometabolic Disease by Modified Macrophages"
This proposal will investigate the role of adiponectin-macrophage axis in the pathogenesis of the Metabolic Syndrome and provide new approaches for preventing and treating cardiometabolic disease.
Role on Project: Principal Investigator

Basic Science Award, Fu (PI) 01/01/12-12/31/14
ADA
"Role of TRIB3 in Glucose-Induced Insulin Resistance"
This proposal will elucidate novel roles for TRIB3 in glucose-induced insulin resistance and in the regulation of fuel metabolism, and develop a new target for therapy in diabetes.
Role on Project: Principal Investigator

R01 PA-10-067, Fu and Garvey (PIs) 09/01/11-08/31/16
NIH
"Role of the Adiponectin-Macrophage Axis in Cardiometabolic Disease"
These studies will provide novel insights into the causes of Metabolic Syndrome, Type 2 Diabetes, and heart disease, which exert a huge burden of patient suffering and social costs, and identify new ways to prevent and treat cardiometabolic disease.
Role on Project: Co-Principal Investigator

Completed Research Support

Research Award, Fu (PI) 04/01/09-03/31/10
UAB Diabetes Research and Training Center
"Analyzing adiponectin receptor transgenic mice for metabolic syndrome"
The major goals of this project are to investigate whether enhanced adiponectin action (i.e., AdipoR1 overexpression) can ameliorate metabolic abnormalities that comprise the metabolic syndrome in mice.
Role on Project: Principal Investigator

Research Award 1-07-RA-49, Fu (PI) 01/01/07-12/31/09
ADA
"Adiponectin Mediated Lipid Metabolism in Atherosclerosis and Metabolic Syndrome"
The major goals of this project are to investigate the mechanisms of preventing metabolic syndrome due to the adiponectin gene overexpression in macrophages.
Role on Project: Principal Investigator

P20 RR016434-010004, Fu (PI) 10/01/01-09/30/06
NIH/NCRR
"Transcriptional programs mediating foam cell formation"
This project is to determine the transcriptional programs involved in both foam cell formation and adipocyte differentiation, and also to identify regulatory elements within the ALBP promoter and generate ALBP transgenic mice.
Role on Project: Principal Investigator

Institutional Research Award, Fu (PI) 01/01/01-12/31/02
MUSC
"ALBP/KLBP: Critical mediators of atherogenic lipid signaling pathways in macrophages"
The aims of this study were to investigate the role of ALBP and KLBP in lipid transport and metabolism in macrophages and to determine the mechanisms involved in oxLDL uptake in macrophages as well as signal transduction parameters involved in the up-regulation of ALBP and KLBP.
Role on Project: Principal Investigator

In: Proteins Researcher Biographical Sketches … ISBN: 978-1-62100-777-7
Editors: H. Z. Wang and M. Tian

Chapter 15

BIOGRAPHICAL SKETCH

NAME:	TITLE:	DATE OF BIRTH:
Katsutoshi Furukawa	**DEPARTMENT OF GERIATRICS AND GERONTOLOGY, TOHOKU UNIVERSITY**	**DEC 1, 1960**

EDUCATION:

Institution and Location	Degree	Year Conferred	Scientific Field
Graduate school Tohoku University	Graduated	1992	
Medical School Yamagata University	Graduated	1988	

CONTACT POINTS:

Address: 1-1 Seiryo-machi Aobaku, Sendai Japan 980-0011

RESEARCH AND PROFESSIONAL EXPERIENCE:

Clinical neurology
Clinical Gerontology
Neuroscience
Neurophysiology
Neuropharmacology

Professional Appointments:

Associate professor, Department of Geriatrics and Gerontology, Tohoku University

Honors:

American Federation for Aging Research, Research Grant, 1998
John A. Hartford Foundation Aging Research Grant Award, 1998
Novartis Foundation for Gerontological Research Award, 2009
Best research award, Japan Geriatrics Society, 2010

Publications during Last Three Years:

(1) Fujiwara, H., Tabuchi, M., Yamaguchi, T., Iwasaki, K., Furukawa, K., Sekiguchi, K., Ikarashi, Y., Kudo, Y., Higuchi, M, Saido, T.C., Maeda, S., Takashima, A., Hara, M., Yaegashi, N., Kase, Y., Arai, H. A traditional medicinal herb Paeonia suffruticosa and its active constituent 1,2,3,4,6-penta-O-galloyl-ï• ¢-D-glucopyranose have potent anti-aggregation effects on Alzheimer's amyloid ï• ¢ proteins in vitro and in vivo. Journal of Neurochemistry 109: 1648-1657, 2009.

(2) Waragai, M., Okamura, N., Furukawa, K., Tashiro, M., Furumoto, S., Funaki, Y., Kato, M., Iwata, R., Yanai, K., Kudo, Y., Arai, H. Comparison study of amyloid PET and voxel-based morphometry analysis in mild cognitive impairment and Alzheimer's disease. Journal of Neurological Sciences 285: 100-108, 2009.

(3) Ohrui T, Yamasaki M, He M, Ebihara S, Ebihara T, Asada M, Yamanda S, Asamura T, Une K, Yoshida M, Kosaka Y, Furukawa K, Arai H. Homicides of disabled older persons by their caregivers and preference for place of death in community-dwelling elderly people in Japan Nippon Ronen Igakkai Zasshi. 2009 Jul;46(4):306-8.

(4) Yamaya M, Yoshida M, Yamasaki M, Kubo H, Furukawa K, Arai H. Seizure and pneumonia in an elderly patient with systemic lupus erythematosus. J Am Geriatr Soc. 2009 Sep;57(9):1709-11.

(5) Furukawa K, Okamura N, Tashiro M, Waragai M, Furumoto S, Iwata R, Yanai K, Kudo Y, Arai H. Amyloid PET in mild cognitive impairment and Alzheimer's disease with BF-227: comparison to FDG-PET. J Neurol. 2010 May;257(5):721-7.

(6) Okamura N, Shiga Y, Furumoto S, Tashiro M, Tsuboi Y, Furukawa K, Yanai K, Iwata R, Arai H, Kudo Y, Itoyama Y, Doh-ura K. In vivo detection of prion amyloid plaques using [(11)C]BF-227 PET. Eur J Nucl Med Mol Imaging. 2010 May;37(5):934-41.

(7) Asamura T, Ohrui T, Une K, Furukawa K, Arai H. Centrally Active ACEIs and cognitive decline. Arch Intern Med. 2010 Jan 11;170(1):107-8

(8) Asamura T, Ohrui T, Nakayama K, He M, Yamasaki M, Ebihara T, Ebihara S, Furukawa K, Arai H. Low serum 1,25-dihydroxyvitamin D level and risk of respiratory infections in institutionalized older people. Gerontology. 2010;56(6):542-3.

(9) Une K, Takei YA, Tomita N, Asamura T, Ohrui T, Furukawa K, Arai H. Adiponectin in plasma and cerebrospinal fluid in MCI and Alzheimer's disease. Eur J Neurol. 2010 Aug 18

(10) Shao H, Okamura N, Sugi K, Furumoto S, Furukawa K, Tashiro M, Iwata R, Matsuda H, Kudo Y, Arai H, Fukuda H, Yanai K. Voxel-based analysis of amyloid positron emission tomography probe [C]BF-227 uptake in mild cognitive impairment and alzheimer's disease. Dement Geriatr Cogn Disord. 2010;30(2):101-11.

(11) Liu D, Pitta M, Lee JH, Ray B, Lahiri DK, Furukawa K, Mughal M, Jiang H, Villarreal J, Cutler RG, Greig NH, Mattson MP. The KATP channel activator diazoxide ameliorates amyloid-Î² and tau pathologies and improves memory in the 3xTgAD mouse model of Alzheimer's disease. J Alzheimers Dis. 2010;22(2):443-57.

(12) Futakawa S, Nara K, Miyajima M, Kuno A, Ito H, Kaji H, Shirotani K, Honda T, Tohyama Y, Hoshi K, Hanzawa Y, Kitazume S, Imamaki R, Furukawa K, Tasaki K, Arai H, Yuasa T, Abe M, Arai H, Narimatsu H, Hashimoto Y. A unique N-glycan on human transferrin in CSF: a possible biomarker for iNPH. In press 2011

In: Proteins Researcher Biographical Sketches … ISBN: 978-1-62100-777-7
Editors: H. Z. Wang and M. Tian

Chapter 16

BIOGRAPHICAL SKETCH

NAME:	**TITLE:**	**DATE OF BIRTH:**
Viviana Girasole	**UNIVERSITY OF MESSINA**	**10\03\1954**

EDUCATION:

Institution and Location	Degree	Year Conferred	Scientific Field
Department of Neuroscience, Psychiatric and Anesthesiological Sciences, University	MD Aggregated Professor		

CONTACT POINTS:

Address: viale Regina Margherita 28 98121 Messina Italy

RESEARCH AND PROFESSIONAL EXPERIENCE:

Intensive Care

Professional Appointments:

MD Associated Professor — MD Aggregated Professor

Publications during Last Three Years:

1) Siracusano L, Girasole V: Carbon monoxide and adiponectin in sepsis. Surgery. 2010 May;147(5):755.
2) Siracusano L, Girasole V: The genetics of malignant hyperthermia and related muscular syndromes. Anesth Analg. 2010 Apr 1;110(4):1241;.
3) Siracusano L, Girasole V: Propofol and cardioprotection against arrhythmias. Anesthesiology. 2009 Aug;111(2):447-8.
4) Siracusano L, Girasole V: Glucose and lipid metabolism in sepsis and endotoxemia. Acta Anaesthesiol Scand. 2009 Mar;53(3):413-4.
5) Siracusano L, Girasole V: Intercellular junctions in sepsis. Crit Care Med. 2008 Feb;36(2):659- 60; author reply 660-1.
6) Siracusano L, Girasole V: Sevoflurane and cardioprotection. Br *J* Anaesth. 2008 Feb;100(2):278; author reply 278-9.

In: Proteins Researcher Biographical Sketches … ISBN: 978-1-62100-777-7
Editors: H. Z. Wang and M. Tian

Chapter 17

BIOGRAPHICAL SKETCH

NAME:	TITLE:	DATE OF BIRTH:
Tatyana S. Godovikova	INSTITUTE OF CHEMICAL BIOLOGY AND FUNDAMENTAL MEDICINE, SIBERIAN BRANCH OF RUSSIAN ACADEMY OF SCIENCES	**29.04.1956**

EDUCATION:

Institution and Location	Degree	Year Conferred	Scientific Field
Institute of Chemical Biology and Fundamental Medicine, Siberian Branch of Russian Academy of Sciences.	Sc.D.	2008	Bioorganic Chemistry
Novosibirsk Institute of Bioorganic Chemistry, Siberian Branch of Russian Academy of Sciences	Ph.D.	1986	Bioorganic Chemistry
Chemistry, Novosibirsk State University, Russia	B.Sc.	1978	

CONTACT POINTS:

Address:
8, Lavrentiev Ave, Novosibirsk, 630090, Russia
Tel/Fax +7-(383)-3635161
E-mail godov@niboch.nsc.ru; t_godovikova@mail.ru

RESEARCH AND PROFESSIONAL EXPERIENCE:

1984-1987: Junior Research Scientist, Novosibirsk Institute of Bioorganic Chemistry, Siberian Branch of Russian Academy of Sciences

Professional Appointments:

2009-Present: Professor of Biochemistry and Bioorganic Chemistry, Department of Natural Sciences, Novosibirsk State University (joint appointment)
2006-Present: Chief Research Scientist, Institute of Chemical Biology and Fundamental Medicine, Siberian Branch of Russian Academy of Sciences
1999-2009: Assistant Professor of Biochemistry and Bioorganic Chemistry, Department of Natural Sciences, Novosibirsk State University (joint appointment)
1996-2006: Senior Research Scientist, Institute of Chemical Biology and Fundamental Medicine, Siberian Branch of Russian Academy of Sciences
1993-1999: Lecturer, Department of Natural Sciences, Novosibirsk State University (joint appointment)
1992-1993: Teaching Assistant, Department of Natural Sciences, Novosibirsk State University (joint appointment)
1987-1996: Research Scientist, Institute of Chemical Biology and Fundamental Medicine, Siberian Branch of Russian Academy of Sciences

Honors:

2009: Certificate of Merit "For Strengthening the Bonds between Education and Science" from Novosibirsk State University and Siberian Branch of Russian Academy of Sciences
2009: Badge of Honor "Golden Sigma" for continuous scientific contributions
2007: Badge of Merit "National Asset"
2007: Badge of Honor "Silver Sigma" for continuous scientific contributions
2006: Award of the State Duma of Russian Federation "For Significant Contribution in Preserving and Enhancing the Intellectual Potential of Russia"
2000-2003: Leading Russian Scientists Fellowship
1999: Certificate of Merit "For Continuous and Fruitful Work in Russian Science" from Russian Academy of Sciences
1998: "Soros High School Teacher" Award
1989: The Lenin's Komsomol Prize in the field of Science and Technology

Publications during Last Three Years:

Popova, T. V.; Reinbolt, J.; Ehresmann, B.; Shakirov, M. M.; Serebriakova, M. V.; Gerassimova, Y. V.; Knorre, D. G.; Godovikova, T. S. (2010). "Why do p-nitro-substituted aryl azides provide unintended dark reactions with proteins?" in J. Photochem. Photobiol. B., 100, 19-29.

Gerasimova, Y. V.; Bobik, T. V.; Ponomarenko, N. A.; Shakirov, M. M.; Zenkova, M. A.; Tamkovich, N. V.; Popova, T. V.; Knorre, D. G.; Godovikova T. S. (2010). "RNA-hydrolyzing activity of human serum albumin and its recombinant analogue" in Bioorg. Med. Chem. Lett., 20, 1427-1431.

Yarinich, L. A.; Koroleva, L. S.; Godovikova, T. S.; Silnikov, V. N. (2010). "Design and synthesis of polyfunctional spacers based on biodegradable peptides" in Proceedings of the 31st European Peptide Symposium, M. Lebl, M. Meldal, K. J. Jensen, T. H.-J. Jensen (Editors), European Peptide Society, Sept. 5-10, 2010, Copenhagen, Denmark, 526-527.

Knorre, D. G.; Kudryashova, N. V.; Godovikova, T. S. (2009). "Chemical and functional aspects of posttranslational proteins modification" in ActaNaturae, 3, 32-56.

Gerasimova, Y. V.; Knorre, D. G.; Shakirov, M. M.; Godovikova T. S. (2008). "Human Serum Albumin as a Catalyst of RNA Cleavage: N-Homocysteinylation and N-Phosphorylation by Oligonucleotide Affinity Reagent Alter the Reactivity of the Protein" in Bioorg. Med. Chem. Lett., 18, 5396-5398.

Gerasimova, Y. V.; Erchenko, I. A.; Shakirov, M. M.; Godovikova, T. S. (2008). "Interaction of Human Serum Albumin and its Clinically Relevant Modification with Oligoribonucleotides" in Bioorg. Med. Chem. Lett., 18, 4511-4514.

In: Proteins Researcher Biographical Sketches ...
Editors: H. Z. Wang and M. Tian
ISBN: 978-1-62100-777-7

Chapter 18

BIOGRAPHICAL SKETCH

NAME:	TITLE:	DATE OF BIRTH:
Naomi Hachiya	TOKYO MEDICAL UNIVERSITY	03/24/1965

EDUCATION:

Institution and Location	Degree	Year Conferred	Scientific Field
Kyushu University	Ph.D.	1995	Molecular Biology
Kyushu University	M.S.		Molecular Biology

CONTACT POINTS:

Address:6-1-1 Shinjuku, Tokyo, Japan 160-8402

RESEARCH AND PROFESSIONAL EXPERIENCE:

Research Fellow of the Japan Society for the Promotion of Science (April 1994-August 1997), Visiting Ph.D.student of Basel university, Switzerland,1994,Post doctoral fellow, University of California, Berkley, Dept. of Molecular and cellular biology(September 1996-August1997)

Professional Appointments:

Associate professor, Department Neurophysiology(2006-present)
Assistant professor, Department Neurophysiology(2005-2006)

Honors:

A prize of Journal of Biochemistry, Japan, in 1994.,
Poster award at a Conference of the 31st Nitoh Foundation in 2003,
The first prize of the international symposium of Prion Disease in 2004.,
Scholarship award of Keystone Symposia in 2005., Outstanding abstract award for GTC Bio,USA in 2007

Publications during Last Three Years:

Possible involvement of calpain-like activity in normal processing of cellular prion proteinNeurosci Lett. 2011 Feb 25;490(2):150-5.,
Detection system based on the conformational change in an aptamer and its application to simple bound/free separation.Biosens Bioelectron. 2009 Jan 1;24(5):1372-6.
Epub 2008 Aug 20.,14-3-3zeta is indispensable for aggregate formation of polyglutamine-expanded huntingtin protein.Neurosci Lett. 2008 Jan 24;431(1):45-50.
Mechanical stress and formation of protein aggregates in neurodegenerative disorders.Med Hypotheses. 2008;70(5):1034-7.

In: Proteins Researcher Biographical Sketches … ISBN: 978-1-62100-777-7
Editors: H. Z. Wang and M. Tian

Chapter 19

BIOGRAPHICAL SKETCH

NAME:	TITLE:	DATE OF BIRTH:
Kuniyuki Hatori	**DEPARTMENT OF BIOSYSTEM ENGINEERING, GRADUATE SCHOOL OF SCIENCE AND TECHNOLOGY, YAMAGATA UNIVERSITY (JAPAN)**	**DECEMBER 28, 1968**

EDUCATION:

Institution and Location	Degree	Year Conferred	Scientific Field
Nagaoka University of Technology (Japan)	Ph. D. (Eng.),	1996	

CONTACT POINTS:

Address: 4-3-16 Jyonan, Yonezawa, Yamagata 992-8510, Japan

RESEARCH AND PROFESSIONAL EXPERIENCE:

Associate Professor at Yamagata University (Japan), 2004 – present

Research Associate at Nagaoka University of Technology (Japan), 1996 – 2004

We research on the molecular movement of actomyosin complex using an in vitro motility assay, in which actin filaments or myosin molecules are directly visualized with fluorescent dyes under a fluorescent microscopy. With the use of a nanometer-accuracy detection of molecular motions, we examine how both the forces supplied from individual myosin molecules and the energies derived from the hydrolysis of ATP molecules are transferred efficiently to movement of an actin filament. The molecular coordinating systems are likely to be useful for fabrication of self-organized structures in the nano-meter scale. Hence, by using the function of motor proteins, we attempt to construct artificial contractile gels and to add the motile activity to artificial plasma membranes.

Publications during Last Three Years:

Hatori, K., Matsui, M., Omote, Y., Slowly modulating fluctuations as mesoscopic distortions occurring on an actin filament, BioSystems 96 (2009) 14-18.

Arii, Y., Hatori, K., Relationship between the flexibility and the motility of actin filaments: Effects of pH, Biochem. Biophys. Res. Commun. 371 (2008) 772-776.

In: Proteins Researcher Biographical Sketches ...
Editors: H. Z. Wang and M. Tian

ISBN: 978-1-62100-777-7

Chapter 20

BIOGRAPHICAL SKETCH

NAME:	TITLE:	DATE OF BIRTH:
Christine F. Hohmann	Ph.D.	**12/14/55**

EDUCATION:

Institution and Location	Degree	Year Conferred	Scientific Field
Brown University, Providence Rhode Island	Ph.D.	1985	Neuroscience

CONTACT POINTS:

Associate Professor, Department of Biology, Morgan State University
1700 East Cold Spring Lane
Baltimore, MD 21251

RESEARCH AND PROFESSIONAL EXPERIENCE:

1980-1984: Thesis Research in the Laboratory of Dr. Ford F. Ebner, Brown University, Providence, R.I.

Dec., 1984: Postdoctoral Fellow, Laboratory of Dr. Joseph T. Coyle, Dept. of Psychiatry, The Johns Hopkins University School of Medicine, Baltimore, MD

April 1989: Instructor, Department of Psychiatry, JHUSM, Baltimore, MD.

July, 1990: Assistant Professor, Division of Child Psychiatry, Department of Psychiatry, JHUSM

Sept. 1991: "Research Scientist, The Kennedy Krieger Research Institute, Baltimore, MD

Sept. 1993: Assistant Professor, Department of Biology, Morgan State University, Baltimore, MD

June 1996: Associate Professor with Tenure, Department of Biology, Morgan State University
September 1997: Graduate Faculty, Morgan State University

Professional Appointments:

1999 – 2002: Panel member: NSF Behavioral and Computational Neuroscience Review Panel;
2001, 2002: Ad hoc, NSF panel review, the Louis Stokes Alliance Program
2003: Ad hoc reviewer, National Association for Autism Research Fellowship Applications;
2001 – 2003: NIGMS panel review, MBRS SCORE (genetics and physiology panel, ad hoc);
2001, 2002: ad hoc member Special Emphasis Panel [SEP] for MBRS RISE and SCORE application
2003 to 2008: Standing Panel member for the MBRS SEP Review Subcommittee, NIGMS;
2003 to 2008: Standing Panel member NIGMS, MBRS SCORE Neuroscience scientific review panel;
1998-present: Advisory Committee Member, NIH PO1 Grant: "Aging of the Brain: Effect of Perinatal Choline Exposure Based at Boston University, P.I. K. Blusztajn (Boston University, School of Medicine)
2000 to 2005: Advisory Board Member, Society for Neuroscience's Minority Fellowship Program
2006-present: Steering Committee Member, CDC training grant 'Research Initiative for Student Enhancement' (P.I. H. Belcher, Kennedy Krieger Institute, Johns Hopkins University)
2005, 2006: fall Ad hoc review panel for NIMH ZMH1-ERB-A SEP CIDAR Mental Health Centers for Schizophrenia and ADHD;
2006 spring: Ad hoc panel review NINDS ZNS1 SRB-M SEP NSRA fellowship review;
2007 spring: Ad hoc proposal review NINDS ZNS1 SRB-M 50, K99, Pathway to Independence;
2009/2010: Ad hoc proposal review ZMH1-ERB-L (Biobehavioral Research Award for Innovative new Scientist)

Honors and Awards:

2002: Deans Award for Excellence in Research, Morgan State University
2002: Fellow of the International Behavioral Neuroscience Society

Publications during Last Three Years:

1) Krasnova, I.N., Betts, E.S., Dada, A., Jefferson, A., Ladenheim, B., Becker, K.G. Cadet, J.L. and Hohmann, C.F. Neonatal dopamine depletion indices change in

morphogenesis and transcriptional programs in the developing cortex: Behavioral implications. Neurotoxicity Research 11(2), 107-130, 2007. 17449454

2) Boylan, C.B., *Blue, M.E., Hohmann, C.F. Modeling early cortical serotonergic deficits in autism. Behav Brain Res 2007; 176:94-108 (*corresponding author). PMCID: 2570481
3) Hohmann, C.F., Walker, E. M., Boylan, C.B., Blue, M.E. Neonatal serotonin depletion alters behavioral responses to spatial change and novelty. Brain Res. 2007; 1139:163-77. PMCID: 1974858
4) Blue, M.E., Johnston, M.V., Moloney, C.B. and Hohmann, C.F. Serotonin dysfunction in autism. In Autism: Current Theories and Evidence. A. Zimmerman Ed. Humana Press, 2008; pgs 111-132.
5) Krasnova, I., Hodges, A., Ladenheim B,, Rhoades, R., Philip, C., Cesena, A., Ivanova, K. and Christine F. Hohmann and Jean Lud Cadet (co-senior authors). Methamphetamine treatment causes delayed decrease in novelty-induced locomotor activity in mice. Neuroscience Research 65 (2009) 160–165.
6) Hohmann C.F. and Blue, M.E. The role of serotonin in cortical development – implications for Autism Spectrum Disorder. In: Handbook of the Behavioral Neurobiology of Serotonin. C.P. Mueller and B. Jacobs, Eds. Academic Press, Elsevier, 2010; pgs 637-665.

In: Proteins Researcher Biographical Sketches …
Editors: H. Z. Wang and M. Tian
ISBN: 978-1-62100-777-7

Chapter 21

BIOGRAPHICAL SKETCH

NAME:	TITLE:	DATE OF BIRTH:
Chuan Hu	PH.D. DEPARTMENT OF BIOCHEMISTRY AND MOLECULAR BIOLOGY, UNIVERSITY OF LOUISVILLE SCHOOL OF MEDICINE, LOUISVILLE, KY 40202	

EDUCATION:

Institution and Location	Degree	Year Conferred	Scientific Field
Columbia University	Ph.D.	2000	

CONTACT POINTS:

Address: 319 Abraham Flexner Way, Room 504, University of Louisville, Louisville, KY 40202

RESEARCH AND PROFESSIONAL EXPERIENCE:

11/2000 – 6/2004 Research Fellow, Cell Biology, Memorial Sloan-Kettering Cancer Center, New York

7/2004 - 11/2004 Postdoctoral Research Scientist, Department of Physiology & Cellular Biophysics, Columbia University Medical Center

Professional Appointments:

12/2004 – 12/2006 Research Assistant Professor, Department of Physiology & Pharmacology, West Virginia University Health Sciences Center

1/2007 – Assistant Professor, Department of Biochemistry & Molecular Biology, University of Louisville School of Medicine

Honors:

2005 Scientist Development Award, the American Heart Association

2007 The Roger H. Herzig Junior Faculty Research Prize, 3rd place, James Graham Brown Cancer Center Annual Retreat, University of Louisville

Publications during Last Three Years:

1. Luftman, K., Hasan, N., Day, P., Hardee, D., and **Hu, C**. 2009. Silencing of VAMP3 inhibits cell migration and integrin-mediated adhesion. *Biochemical and Biophysical Research Communications* 380:65-70.
2. Hasan, N., and **Hu, C.** 2010. Vesicle-associated membrane protein 2 mediates trafficking of alpha5beta1 integrin to the plasma membrane. *Experimental Cell Research* 316:12-23.
3. Hasan, N., Corbin, D., and **Hu, C.** 2010. Fusogenic pairings of vesicle-associated membrane proteins (VAMPs) and plasma membrane t-SNAREs – VAMP5 as the exception. *PLoS One*. 5(12):e14238

In: Proteins Researcher Biographical Sketches ...
Editors: H. Z. Wang and M. Tian

ISBN: 978-1-62100-777-7

Chapter 22

BIOGRAPHICAL SKETCH

NAME:	TITLE:	DATE OF BIRTH:
Miki Imanishi	INSTITUTE FOR CHEMICAL RESEARCH, KYOTO UNIVERSITY, JAPAN	1974/9/20

EDUCATION:

Institution and Location	Degree	Year Conferred	Scientific Field
Kyoto University	B.S.	1997	
Kyoto University	PhD	2002	Pharmaceutical Sciences

CONTACT POINTS:

Address: Gokasho, Uji, Kyoto 61-0011, Japan

RESEARCH AND PROFESSIONAL EXPERIENCE:

2002-2003 Postdoctoral fellow in UCSF

Professional Appointments:

2003- Assistant Professor in Kyoto University

Publications during Last Three Years:

1. Imanishi M, Imamura C, Higashi C, Yan W, Negi S, Futaki S, Sugiura Y. Zinc finger-zinc finger interaction between the transcription factors, GATA-1 and Sp1. Biochem Biophys Res Commun. 2010; 400(4):625-30.
2. Imanishi M, Nakaya T, Morisaki T, Noshiro D, Futaki S, Sugiura Y. Metal-stimulated regulation of transcription by an artificial zinc-finger protein. Chembiochem. 2010; 11(12):1653-5.
3. Imanishi M, Negi S, Sugiura Y. Non-FokI-based zinc finger nucleases. Methods Mol Biol. 2010;649:337-49.
4. Imanishi M, Nakamura A, Morisaki T, Futaki S. Positive and negative cooperativity of modularly assembled zinc fingers. Biochem Biophys Res Commun. 2009; 387(3):440-3.
5. Azuma Y, Imanishi M, Yoshimura T, Kawabata T, Futaki S. Cobalt(II)-responsive DNA binding of a GCN4-bZIP protein containing cysteine residues functionalized with iminodiacetic acid. Angew Chem Int Ed Engl. 2009;48(37):6853-6.
6. Dhanasekaran M, Negi S, Imanishi M, Suzuki M, Sugiura Y. Effects of bulkiness and hydrophobicity of an aliphatic amino acid in the recognition helix of the GAGA zinc finger on the stability of the hydrophobic core and DNA binding affinity. Biochemistry. 2008; 47(45):11717-24.
7. Morisaki T, Imanishi M, Futaki S, Sugiura Y. Rapid transcriptional activity in vivo and slow DNA binding in vitro by an artificial multi-zinc finger protein. Biochemistry. 2008; 47(38):10171-7.
8. Negi S, Imanishi M, Matsumoto M, Sugiura Y. New redesigned zinc-finger proteins: design strategy and its application. Chemistry. 2008;14(11):3236-49. Review.

In: Proteins Researcher Biographical Sketches … ISBN: 978-1-62100-777-7
Editors: H. Z. Wang and M. Tian © 2012 Nova Science Publishers, Inc.

Chapter 23

BIOGRAPHICAL SKETCH

NAME:	TITLE:	DATE OF BIRTH:
Marc G. Jeschke	**MD. PhD. Department of Biochemistry and Molecular Biology, The University of Texas Medical Branch, Galveston, Texas, USA and Sunnybrook Health Sciences Centre, Department of Surgery and Plastic Surgery, University of Toronto, Toronto, Canada**	**1/2/67**

EDUCATION:

Institution and Location	Degree	Year Conferred	Scientific Field
	MD, MMS, PhD		

CONTACT POINTS:

Address: Department of Surgery and Plastic Surgery, University of Toronto, Sunnybrook Health Sciences Centre, 2075 Bayview Avenue, Rm D704, Toronto, Canada M4N 3M5

RESEARCH AND PROFESSIONAL EXPERIENCE:

In 1994, Dr. Jeschke completed his medical school and his thesis with summa cum laude. He completed a research Fellowship from 1996-1999 with the University Texas Medical Branch and Shriners Hospital for Children. During this time he completed his Masters of Medical Science. He returned to Germany, to the Department of Surgery at the University of Regensburg where received his surgical training and passed his General Surgery Boards in 2003, but also was awarded the habilitation (PhD) in Experimental Surgery in 2001. Dr. Jeschke returned to the University Texas Medical Branch and Shriners Hospital for Children in 2004 as Faculty, and was made Burn Attending and Coordinator of Research in 2006. He was named Annie Laurie Howard Chair of Burn Surgery in 2008.

Furthering his interest in burn and critical care, Dr. Jeschke undertook a clinical fellowship to obtain highly specialized training. From August 2005 to June 2006 he was a Burn and Critical Care Fellow at the University Texas Medical Branch and Shriners Hospital for Children, the only program to be certified by ACMG.

In May 2010, Dr. Jeschke was appointed Director of the Ross Tilley Burn Centre at Sunnybrook Health Science Centre. He was also appointed as Surgeon-Scientist in the Division of Plastic and Reconstructive Surgery, Department of Surgery at the University of Toronto. Dr. Jeschke is an Associate Professor of Surgery at the University of Toronto.

Professional Appointments:

Associate Professor of Surgery, Director of the Ross Tilley Burn Centre, Surgeon-Scientist

Honors:

Dr.. Jeschke has won numerous awards for teaching and research, including the prestigious Fellow Award of the American Surgical Association Foundation (2006), First Prize Poster Presentation, American Burn Association (2005), Surgery Specialty Award, Society Critical Care Medicine (2004), Von-Langenbeck-Prize from the German College of Surgeons (2004), Prize fort the best scientific work from the German College of Surgeons, Critical Care Chapter (2003), Dr. Werner Fekl FÃ rderprize of the German Society of Critical Care Medicine (2002), Best scientific presentation at the 1st World meeting of the Surgical Infection Society, Madrid, Spain (2002), Wound healing-Prize, German College of Surgeons (2001), Otto-GÃtze Prize of the Bavarian Society of Surgery (2001), Johann-Nepomuk-von-Nussbaum Prize der Bayerischen Gesellschaft fÃ¼r Chirurgie (2000), and American College of Surgeons Committee on Trauma, Region VI. Winner Residents Award and selected for presentation at the National Competition, Washington DC (1998).

Publications during Last Three Years:

Burns: where are we standing with propranolol, oxandrolone, recombinant human growth hormone, and the new incretin analogs?. [Review]

Gauglitz GG. Williams FN. Herndon DN. Jeschke MG. Current Opinion in Clinical Nutrition & Metabolic Care. 14(2):176-81, 2011 Mar.

Propranolol decreases cardiac work in a dose-dependent manner in severely burned children. Williams FN. Herndon DN. Kulp GA. Jeschke MG. Surgery. 149(2):231-9, 2011 Feb.

Impact of anesthesia, analgesia, and euthanasia technique on the inflammatory cytokine profile in a rodent model of severe burn injury. Al-Mousawi AM. Kulp GA. Branski LK. Kraft R. Mecott GA. Williams FN. Herndon DN. Jeschke MG. Shock. 34(3):261-8, 2010 Sep.

The use of exenatide in severely burned pediatric patients. Mecott GA. Herndon DN. Kulp GA. Brooks NC. Al-Mousawi AM. Kraft R. Rivero HG. Williams FN. Branski LK. Jeschke MG. Critical Care (London, England). 14(4):R153, 2010.

Invited commentary on "The demographics of modern burn care: should most burns be cared for by non-burn surgeons?". Al-Mousawi AM. Jeschke MG. Herndon DN. American Journal of Surgery. 201(1):97-9, 2011 Jan.

M2b monocytes predominated in peripheral blood of severely burned patients. Kobayashi M. Jeschke MG. Shigematsu K. Asai A. Yoshida S. Herndon DN. Suzuki F. Journal of Immunology. 185(12):7174-9, 2010 Dec 15.

Plasma proteome response to severe burn injury revealed by 18O-labeled "universal" reference-based quantitative proteomics. Qian WJ. Petritis BO. Kaushal A. Finnerty CC. Jeschke MG. Monroe ME. Moore RJ. Schepmoes AA. Xiao W. Moldawer LL. Davis RW. Tompkins RG. Herndon DN. Camp DG 2nd. Smith RD. Inflammation and the Host Response to Injury Large Scale Collaborative Research Program. Journal of Proteome Research. 9(9):4779-89, 2010 Sep 3.

What's new in Shock, October 2010?. Shahrokhi S. Jeschke MG. Shock. 34(4):323-6, 2010 Oct.

Tocopherol adipose tissue stores are depleted after burn injury in pediatric patients. Traber MG. Leonard SW. Traber DL. Traber LD. Gallagher J. Bobe G. Jeschke MG. Finnerty CC. Herndon D. American Journal of Clinical Nutrition. 92(6):1378-84, 2010 Dec.

Intensive insulin therapy in severely burned pediatric patients: a prospective randomized trial. Jeschke MG. Kulp GA. Kraft R. Finnerty CC. Mlcak R. Lee JO. Herndon DN. American Journal of Respiratory & Critical Care Medicine. 182(3):351-9, 2010 Aug 1.

Extent and magnitude of catecholamine surge in pediatric burned patients. Kulp GA. Herndon DN. Lee JO. Suman OE. Jeschke MG. Shock. 33(4):369-74, 2010 Apr.

Beta blockade: the right time, the right dose, the right receptor!. Al-Mousawi AM. Jeschke MG. Herndon DN. Critical Care Medicine. 38(2):688-9, 2010 Feb.

Glucose control in severely thermally injured pediatric patients: what glucose range should be the target?. Jeschke MG. Kraft R. Emdad F. Kulp GA. Williams FN. Herndon DN. Annals of Surgery. 252(3):521-7; discussion 527-8, 2010 Sep.

Analysis of factorial time-course microarrays with application to a clinical study of burn injury. Zhou B. Xu W. Herndon D. Tompkins R. Davis R. Xiao W. Wong WH. Inflammation and Host Response to Injury Program. Toner M. Warren HS. Schoenfeld

DA. Rahme L. McDonald-Smith GP. Hayden D. Mason P. Fagan S. Yu YM. Cobb JP. Remick DG. Mannick JA. Lederer JA. Gamelli RL. Silver GM. West MA. Shapiro MB. Smith R. Camp DG 2nd. Qian W. Storey J. Mindrinos M. Tibshirani R. Lowry S. Calvano S. Chaudry I. West MA. Cohen M. Moore EE. Johnson J. Moldawer LL. Baker HV. Efron PA. Balis UG. Billiar TR. Ochoa JB. Sperry JL. Miller-Graziano CL. De AK. Bankey PE. Finnerty CC. Jeschke MG. Minei JP. Arnoldo BD. Hunt JL. Horton J. Cobb JP. Brownstein B. Freeman B. Maier RV. Nathens AB. Cuschieri J. Gibran N. Klein M. O'Keefe G. Proceedings of the National Academy of Sciences of the United States of America. 107(22):9923-8, 2010 Jun 1.

Pre-clinical evaluation of liposomal gene transfer to improve dermal and epidermal regeneration. Branski LK. Masters OE. Herndon DN. Mittermayr R. Redl H. Traber DL. Cox RA. Kita K. Jeschke MG. Gene Therapy. 17(6):770-8, 2010 Jun.

What's new in Shock, June 2010?. Al-Mousawi AM. Jeschke MG. Herndon DN. Shock. 33(6):559-61, 2010 Jun.

Isolation and characterization of mesenchymal stem cells from the sub-amniotic human umbilical cord lining membrane. Kita K. Gauglitz GG. Phan TT. Herndon DN. Jeschke MG. Stem Cells & Development. 19(4):491-502, 2010 Apr.

Neonate twin with staphylococcal scalded skin syndrome from a renal source. Norbury WB. Gallagher JJ. Herndon DN. Branski LK. Oehring PE. Jeschke MG. Pediatric Critical Care Medicine. 11(2):e20-3, 2010 Mar.

Measurement of body composition in burned children: is there a gold standard?. Branski LK. Norbury WB. Herndon DN. Chinkes DL. Cochran A. Suman O. Benjamin D. Jeschke MG. Jpen: Journal of Parenteral & Enteral Nutrition. 34(1):55-63, 2010 Jan-Feb.

Insulin increases resistance to burn wound infection-associated sepsis. Gauglitz GG. Toliver-Kinsky TE. Williams FN. Song J. Cui W. Herndon DN. Jeschke MG. Critical Care Medicine. 38(1):202-8, 2010 Jan.

The role of hyperglycemia in burned patients: evidence-based studies. [Review] [124 refs] Mecott GA. Al-Mousawi AM. Gauglitz GG. Herndon DN. Jeschke MG. Shock. 33(1):5-13, 2010 Jan.

The hypermetabolic response to burn injury and interventions to modify this response. [Review] [138 refs] Williams FN. Herndon DN. Jeschke MG. Clinics in Plastic Surgery. 36(4):583-96, 2009 Oct.

Burn teams and burn centers: the importance of a comprehensive team approach to burn care. [Review] [34 refs] Al-Mousawi AM. Mecott-Rivera GA. Jeschke MG. Herndon DN. Clinics in Plastic Surgery. 36(4):547-54, 2009 Oct.

Cytokine expression profile over time in burned mice. Finnerty CC. Przkora R. Herndon DN. Jeschke MG. Cytokine. 45(1):20-5, 2009 Jan.

Calcium and ER stress mediate hepatic apoptosis after burn injury. Jeschke MG. Gauglitz GG. Song J. Kulp GA. Finnerty CC. Cox RA. Barral JM. Herndon DN. Boehning D. Journal of Cellular & Molecular Medicine. 13(8B):1857-65, 2009 Aug.

The leading causes of death after burn injury in a single pediatric burn center. Williams FN. Herndon DN. Hawkins HK. Lee JO. Cox RA. Kulp GA. Finnerty CC. Chinkes DL. Jeschke MG. Critical Care (London, England). 13(6):R183, 2009.

Hydrogen sulfide is an endogenous stimulator of angiogenesis. Papapetropoulos A. Pyriochou A. Altaany Z. Yang G. Marazioti A. Zhou Z. Jeschke MG. Branski LK. Herndon DN.

Wang R. Szabo C. Proceedings of the National Academy of Sciences of the United States of America. 106(51):21972-7, 2009 Dec 22.

Emerging infections in burns. [Review] [70 refs] Branski LK. Al-Mousawi A. Rivero H. Jeschke MG. Sanford AP. Herndon DN. Surgical Infections. 10(5):389-97, 2009 Oct.

The hepatic response to thermal injury: is the liver important for postburn outcomes?. [Review] [143 refs] Jeschke MG. Molecular Medicine. 15(9-10):337-51, 2009 Sep-Oct.

Severe burn-induced endoplasmic reticulum stress and hepatic damage in mice. Song J. Finnerty CC. Herndon DN. Boehning D. Jeschke MG. Molecular Medicine. 15(9-10):316-20, 2009 Sep-Oct.

Pathophysiology of burns. [Review] [94 refs] Keck M. Herndon DH. Kamolz LP. Frey M. Jeschke MG. Wiener Medizinische Wochenschrift. 159(13-14):327-36, 2009.

What's new in Shock, August 2009?. Jeschke MG. Herndon DN. Shock. 32(2):119-21, 2009 Aug.

A genomic score prognostic of outcome in trauma patients. Warren HS. Elson CM. Hayden DL. Schoenfeld DA. Cobb JP. Maier RV. Moldawer LL. Moore EE. Harbrecht BG. Pelak K. Cuschieri J. Herndon DN. Jeschke MG. Finnerty CC. Brownstein BH. Hennessy L. Mason PH. Tompkins RG. Inflammation and Host Response to Injury Large Scale Collaborative Research Program. Molecular Medicine. 15(7-8):220-7, 2009 Jul-Aug.

Starvation-induced proximal gut mucosal atrophy diminished with aging. Song J. Wolf SE. Wu XW. Finnerty CC. Gauglitz GG. Herndon DN. Jeschke MG. Jpen: Journal of Parenteral & Enteral Nutrition. 33(4):411-6, 2009 Jul-Aug.

Modulation of the hypermetabolic response to trauma: temperature, nutrition, and drugs. Williams FN. Jeschke MG. Chinkes DL. Suman OE. Branski LK. Herndon DN. Journal of the American College of Surgeons. 208(4):489-502, 2009 Apr.

Hepatic gene expression during endotoxemia. Croner RS. Hohenberger W. Jeschke MG. Journal of Surgical Research. 154(1):126-34, 2009 Jun 1.

Abnormal insulin sensitivity persists up to three years in pediatric patients post-burn. Gauglitz GG. Herndon DN. Kulp GA. Meyer WJ 3rd. Jeschke MG. Journal of Clinical Endocrinology & Metabolism. 94(5):1656-64, 2009 May.

Acute propranolol infusion stimulates protein synthesis in rabbit skin wound. Zhang XJ. Meng C. Chinkes DL. Finnerty CC. Aarsland A. Jeschke MG. Herndon DN. Surgery. 145(5):558-67, 2009 May.

A review of gene and stem cell therapy in cutaneous wound healing. [Review] [88 refs] Branski LK. Gauglitz GG. Herndon DN. Jeschke MG. Burns. 35(2):171-80, 2009 Mar.

Large-scale multiplexed quantitative discovery proteomics enabled by the use of an (18)O-labeled "universal" reference sample. Qian WJ. Liu T. Petyuk VA. Gritsenko MA. Petritis BO. Polpitiya AD. Kaushal A. Xiao W. Finnerty CC. Jeschke MG. Jaitly N. Monroe ME. Moore RJ. Moldawer LL. Davis RW. Tompkins RG. Herndon DN. Camp DG. Smith RD. Inflammation and the Host Response to Injury Large Scale Collaborative Research Program. Journal of Proteome Research. 8(1):290-9, 2009 Jan.

A porcine model of full-thickness burn, excision and skin autografting. Branski LK. Mittermayr R. Herndon DN. Norbury WB. Masters OE. Hofmann M. Traber DL. Redl H. Jeschke MG. Burns. 34(8):1119-27, 2008 Dec.

Impact of oxandrolone treatment on acute outcomes after severe burn injury. Pham TN. Klein MB. Gibran NS. Arnoldo BD. Gamelli RL. Silver GM. Jeschke MG. Finnerty CC.

Tompkins RG. Herndon DN. Journal of Burn Care & Research. 29(6):902-6, 2008 Nov-Dec.

Characterization of the inflammatory response during acute and post-acute phases after severe burn. Gauglitz GG. Song J. Herndon DN. Finnerty CC. Boehning D. Barral JM. Jeschke MG. Shock. 30(5):503-7, 2008 Nov.

Temporal cytokine profiles in severely burned patients: a comparison of adults and children.

Finnerty CC. Jeschke MG. Herndon DN. Gamelli R. Gibran N. Klein M. Silver G. Arnoldo B. Remick D. Tompkins RG. Investigators of the Inflammation and the Host Response Glue Grant. Molecular Medicine. 14(9-10):553-60, 2008 Sep-Oct.

Pathophysiologic response to severe burn injury. Jeschke MG. Chinkes DL. Finnerty CC. Kulp G. Suman OE. Norbury WB. Branski LK. Gauglitz GG. Mlcak RP. Herndon DN. Annals of Surgery. 248(3):387-401, 2008 Sep.

Topical management of facial burns. [Review] [69 refs] Leon-Villapalos J. Jeschke MG. Herndon DN. Burns. 34(7):903-11, 2008 Nov.

Insulin resistance postburn: underlying mechanisms and current therapeutic strategies. [Review] [154 refs] Gauglitz GG. Herndon DN. Jeschke MG. Journal of Burn Care & Research. 29(5):683-94, 2008 Sep-Oct.

Are serum cytokines early predictors for the outcome of burn patients with inhalation injuries who do not survive?. Gauglitz GG. Finnerty CC. Herndon DN. Mlcak RP. Jeschke MG. Critical Care (London, England). 12(3):R81, 2008.

Second hit post burn increased proximal gut mucosa epithelial cells damage. Song J. Wolf SE. Herndon DN. Wu XW. Jeschke MG. Shock. 30(2):184-8, 2008 Aug.

Gender differences in pediatric burn patients: does it make a difference?. Jeschke MG. Mlcak RP. Finnerty CC. Norbury WB. Przkora R. Kulp GA. Gauglitz GG. Zhang XJ. Herndon DN. Annals of Surgery. 248(1):126-36, 2008 Jul.

Combination of recombinant human growth hormone and propranolol decreases hypermetabolism and inflammation in severely burned children. Jeschke MG. Finnerty CC. Kulp GA. Przkora R. Mlcak RP. Herndon DN. Pediatric Critical Care Medicine. 9(2):209-16, 2008 Mar. 58. Urinary cortisol and catecholamine excretion after burn injury in children. Norbury WB. Herndon DN. Branski LK. Chinkes DL. Jeschke MG. Journal of Clinical Endocrinology & Metabolism. 93(4):1270-5, 2008 Apr.

Upper airway mucus deposition in lung tissue of burn trauma victims. Cox RA. Mlcak RP. Chinkes DL. Jacob S. Enkhbaatar P. Jaso J. Parish LP. Traber DL. Jeschke MG. Herndon DN. Hawkins HK. Shock. 29(3):356-61, 2008 Mar.

Amnion in the treatment of pediatric partial-thickness facial burns. Branski LK. Herndon DN. Celis MM. Norbury WB. Masters OE. Jeschke MG. Burns. 34(3):393-9, 2008 May.

Age differences in inflammatory and hypermetabolic postburn responses. Jeschke MG. Norbury WB. Finnerty CC. Mlcak RP. Kulp GA. Branski LK. Gauglitz GG. Herndon B. Swick A. Herndon DN. Pediatrics. 121(3):497-507, 2008 Mar.

The factor age and the recovery of severely burned children. Przkora R. Herndon DN. Jeschke MG. Burns. 34(1):41-4, 2008 Feb.

Insulin decreases inflammatory signal transcription factor expression in primary human liver cells after LPS challenge. Jeschke MG. Klein D. Thasler WE. Bolder U. Schlitt HJ. Jauch KW. Weiss TS. Molecular Medicine. 14(1-2):11-9, 2008 Jan-Feb.

Suicidal intention and self-immolation: what is the outcome?. Norbury WB. Jeschke MG. Herndon DN. Critical Care Medicine. 36(1):345-6, 2008 Jan.

Plasma proteome response to severe burn injury revealed by 18O-labeled "universal" reference-based quantitative proteomics. Qian WJ. Petritis BO. Kaushal A. Finnerty CC. Jeschke MG. Monroe ME. Moore RJ. Schepmoes AA. Xiao W. Moldawer LL. Davis RW. Tompkins RG. Herndon DN. Camp DG 2nd. Smith RD. Inflammation and the Host Response to Injury Large Scale Collaborative Research Program. Journal of Proteome Research. 9(9):4779-89, 2010 Sep 3.

Large-scale multiplexed quantitative discovery proteomics enabled by the use of an (18)O-labeled "universal" reference sample. Qian WJ. Liu T. Petyuk VA. Gritsenko MA. Petritis BO. Polpitiya AD. Kaushal A. Xiao W. Finnerty CC. Jeschke MG. Jaitly N. Monroe ME. Moore RJ. Moldawer LL. Davis RW. Tompkins RG. Herndon DN. Camp DG. Smith RD. Inflammation and the Host Response to Injury Large Scale Collaborative Research Program. Journal of Proteome Research. 8(1):290-9, 2009 Jan. -

In: Proteins Researcher Biographical Sketches …
Editors: H. Z. Wang and M. Tian
ISBN: 978-1-62100-777-7

Chapter 24

BIOGRAPHICAL SKETCH

NAME:	TITLE:	DATE OF BIRTH:
Marina G. Kalyuzhnaya	**UNIVERSITY OF WASHINGTON**	**06/20/1972**

EDUCATION:

Institution and Location	Degree	Year Conferred	Scientific Field
Center for Microbiology and Biotechnology//Institute of Biochemistry and Physiology of Microorganisms, Russian Academy of Sciences, Russia	*PhD.*	**2000**	*Microbiology*
Dnepropetrovsk State University, Department of Microbiology, Ukraine	*BS/ MS*	**1994**	*Biology, Microbiology*

CONTACT POINTS:

Address: University of Washington, Department of Microbiology, Box 355014, Seattle WA, 98105

Professional Appointments:

2006- present: Research Assistant Professor, Department of Microbiology, University of Washington

2001–2006; Research Associate, Department of Chemical Engineering, University of Washington

1997–2002; Junior Research Scientist, G.K. Skryabin Institute of Biochemistry & Physiology of Microorganisms, Russian Academy of Sciences

1995–1997: Engineer-investigator, G.K. Skryabin Institute of Biochemistry & Physiology of Microorganisms, Russian Academy of Sciences

Publications during Last Three Years:

[1] Konopka, M., Strovas, T. J., Ojala, D.S., Chistoserdova, L., Lidstrom, M.E., & *M.G.* Kalyuzhnaya. 2011. Respiration response imaging for real time detection of microbial function at the single cell level. Appl. Environ. Microbiol. 77 (1): 67-72

[2] Tavormina P.L., V. J. Orphan, M.G. Kalyuzhnaya, M.S. M. Jetten & M.G. Klotz. 2011. A novel family of functional operons encoding methane/ammonia monooxygenase-related proteins in gammaproteobacterial methanotrophs. Environmental Microbiology Reports. 3 (1): 91–100

[3] Strovas T.J., S.C. McQuaide, J.B. Anderson, V. Nandakumar, M. G. Kalyuzhnaya, L.W. Burgess, M.R. Holl, D.R. Meldrum, & M. E. Lidstrom. 2010. Direct measurement of oxygen consumption rates from attached and unattached cells in a reversibly sealed, diffusionally isolated sample chamber. Advances in Bioscience and Biotechnology. ABB. 1: 398-408mi

[4] 4Stein L.Y., Yoon S., Semrau J.D., Dispirito A.A., Murrell J.C., Vuilleumier S., Kalyuzhnaya M.G., Op den Camp H.J., Bringel F., Bruce D., Cheng J.F., Copeland A., Goodwin L., Han S., Hauser L., Jetten M.S., Lajus A., Land M.L., Lapidus A., Lucas S., Médigue C., Pitluck S., Woyke T., Zeytun A., Klotz M.G. 2010. Genome sequence of the obligate methanotroph, *Methylosinus trichosporium* strain OB3b. J Bacteriol. 192(24):6497-8

[5] Latypova, E., Y.S. Wang, T. Wang, M. Hackett, H. Schafer, & M. G. Kalyuzhnaya. 2010. Genetics of the amino-acid mediated methylamine utilization pathway in the facultative methylotrophic beta-proteobacterium *Methyloversatilis universalis* FAM5. Mol.Microbiol. 75(2): **426-39**

[6] Mustakhimov I. I., A. S. Reshetnikov, A. S. Glukhov, V. N. Khmelenina, M. G. Kalyuzhnaya, & Y. A. Trotsenko. 2010. Identification and characterization of EctR, a new transcriptional regulator of the ectoine biosynthesis genes in the halotolerant methanotroph *Methylomicrobium alcaliphilum* 20Z. J. Bacteriology, 192(2): 410-7

[7] Kalyuzhnaya, M.G., Martens-Habbena, W., Wang, T., Stolyar, S.M., Hackett, M., Stahl, D.A., Lidstrom, M.E. & Chistoserdova, L. 2009. Microbial populations in freshwater lake sediment linking denitrification to methanol oxidation as assessed by stable isotope probing and pure culture analysis. Environ. Microbiol. Reports. 1(5): 385–92

[8] Kalyuzhnaya, M.G., Beck, D.A.C., Suciu, D., Pozhitkov, A., M.E. Lidstrom & L. Chistoserdova. 2009. Functioning *in situ*: gene expression in *Methylotenera mobilis* in its native environment as assessed through transcriptomics. The ISME Journal. 4: 388–398

[9] Bosch G., Kalyuzhnaya M.G., Wang T., Latypova E., Hackett M., & L. Chistoserdova. 2009. Metagenome based proteomic analysis of *Methylotenera mobilis*. Microbiology, 55:1103-10.

[10] Kalyuzhnaya M.G., Lapidus A., Ivanova N., Copeland A.C., McHardy A.C., Szeto E. Salamov A., Grigoriev I.V., Suciu D., Levine S.R., Markowitz V.M., Rigoutsos I., Tringe S.G., Bruce D.C., Richardson P.M., Lidstrom M.E., & L. Chistoserdova 2008. High-resolution metagenomics targets specific functional types in complex microbial communities. Nature Biotechnology. 26: 1029 - 34

[11] Ojala D.S., Beck , D.A.C. & Kalyuzhnaya, M. G.. Genetic systems for moderately halo(alkali)philic bacteria of the genus *Methylomicrobium*. Methods Enzymol. 495: 99-118.

[12] Konopka, M.C., McQuaide, S., Ojala, D.S., Kalyuzhnaya, M.G., & M.E. Lidstrom. 2011. Single cell methods for methane oxidation analysis. Methods Enzymol. 495:149-66.

[13] Kalyuzhnaya, M.G., Beck, D.A.C., & L. Chistoserdova. Functional metagenomics of methylotrophs. Methods Enzymol. 495: 81-98.

[14] Chistoserdova, L., Kalyuzhnaya M.G., & M.E. Lidstrom. 2009. The expanding world of methylotrophic metabolism. Annu Rev Microbiol. 63:477-99

In: Proteins Researcher Biographical Sketches …
Editors: H. Z. Wang and M. Tian
ISBN: 978-1-62100-777-7

Chapter 25

BIOGRAPHICAL SKETCH

NAME:	TITLE:	DATE OF BIRTH:
Ippei Kanazawa	SHIMANE UNIVERSITY FACULTY OF MEDICINE	Nov 19, 1974

EDUCATION:

Institution and Location	Degree	Year Conferred	Scientific Field
Shimane University Graduate School of Medicine	Ph.D	2009	Medical Science
Shimane University School of Medicine	M.D.	2001	

CONTACT POINTS:

Address: 89-1 Enya-cho, Izumo, Shimane 693-8501, Japan

RESEARCH AND PROFESSIONAL EXPERIENCE:

2009-present: Postdoctoral fellow in Calcium Research Laboratory, McGill University, Montreal, Quebec, Canada
Mentor and research supervisor: Geoffrey N. Hendy, Ph.D.

2007-2009: Assistant Professor of Medicine in Division of Internal Medicine 1, Shimane University Faculty of Medicine, Izumo, Shimane, Japan
Mentor and research supervisor: Toshitsugu Sugimoto, M.D., Ph.D.

2004-200: Senior Resident in Endocrinology, Metabolism, and Internal medicine, Shimane University Faculty of Medicine, Izumo, Shimane, Japan
Clinical Mentors: Yuzuru Kato, M.D., Ph.D. and Toshitsugu Sugimoto, M.D., Ph.D.

2003-2004: Resident in Internal Medicine, Ochi Hospital, Ochi, Shimane, Japan
Program Director: Yuzuru Kato, M.D., Ph.D.
Clinical Mentor: Kazunari Sota, M.D.
2001.12-2003: Resident in Endocrinology, Metabolism, and Internal Medicine, Masuda Red Cross Hospital, Masuda, Shimane, Japan
Program Director: Yuzuru Kato, M.D., Ph.D.
Clinical Mentor: Motoi Sohmiya, M.D., Ph.D.
2001.4-2001.11: Resident in Internal Medicine 1, Shimane University Faculty of Medicine, Izumo, Shimane, Japan
Program Director: Yuzuru Kato, M.D., Ph.D.
Clinical Mentor: Toshiaki Mori, M.D., Ph.D.

Honors:

2011: Travel award on Annual McGill University Endocrine Retreat 2011
2011: Post-Doctoral Research Fellowships from Government of Canada 2011-2012
2011: Postdoctoral training award renewal from Fonds de la recherché en santé Quebec
2010: Abroad research fellowship from Uehara Memorial Foundation
2010: Abroad research award from Kanae Foundation for the promotion of medical science
2009: Postdoctoral training award from Fonds de la recherché en santé Quebec
2007: Research exhortation award from Japan Osteoporosis Society
2007: Plenary research award from Japan Endocrine Society
2006: Academic exhortation award from Japan Osteoporosis Society
2006: Asia travel award from Japanese Society for Bone and Mineral Research

Publications during Last Three Years:

Kanazawa I*, Yamaguchi T, Sugimoto T. Effects of intensive glycemic control on serum levels of Insulin-like growth factor-I and dehydroepiandrosterone sulfate in type 2 diabetes mellitus. Journal of Endocrinological Investigation, in press.

Yamaguchi T, Yamamoto M, Kanazawa I, Yamauchi M, Yano S, Tanaka N, Nitta E, Fukuma A, Uno S, Sho-no T, Sugimoto T. Quantitative ultrasound and vertebral fractures in patients with type 2 diabetes. Journal of Bone and Mineral Metabolism, Epub ahead of print.

Kanazawa I*, Yamaguchi T, Sugimoto T. Relationship between bone biochemical markers versus glucose/lipid metabolism and atherosclerosis; a longitudinal study in type 2 diabetes mellitus. Diabetes Research and Clinical Practice, Epub ahead of print.

Kanazawa I*, Yamamoto M, Yamaguchi T, Sugimoto T. Effects of metformin and pioglitazone on serum pentosidine levels in type 2 diabetes mellitus. Experimental and Clinical Endocrinology & Diabetes, Epub ahead of print.

Yamaguchi T, Kanazawa I, Takaoka S, Sugimoto T. Serum calcium is positively correlated with fasting plasma glucose and insulin resistance, independent of parathyroid hormone, in male patients with type 2 diabetes mellitus. Metabolism Clinical and Experimental, Epub ahead of print.

Kanazawa I*, Yano S, Notsu Y, Yamaguchi T, Nabika T, Sugimoto T. Asymmetric dimethylarginine as a risk factor for cardiovascular disease in Japanese patients with type 2 diabetes mellitus. Clinical Endocrinology, 2011 Apr; 74(4): 467-72.

Kanazawa I*, Yamaguchi T, Tada Y, Yamauchi M, Yano S, Sugimoto T. Serum osteocalcin level is positively associated with insulin sensitivity and secretion in patients with type 2 diabetes. Bone, 2011 Apr; 48(4): 270-5.

Kanazawa I*, Yamaguchi T, Sugimoto T. Serum Insulin-like growth factor-I is a maker for assessing the severity of vertebral fractures in postmenopausal women with type 2 diabetes mellitus. Osteoporosis International, 2011 Apr; 22(4): 1191-8.

Kanazawa I*, Yamaguchi T, Yamauchi M, Yamamoto M, Kurioka S, Yano S, Sugimoto T. Serum undercarboxylated osteocalcin was inversely associated with plasma glucose level and fat mass in type 2 diabetes mellitus. Osteoporosis International, 2011 Jan; 22 (1): 187-94.

Kanazawa I*, Yano S, Yamaguchi T, Notsu Y, Nabika T, Sugimoto T. Relationships between dimethylarginine and the presence of vertebral fractures in type 2 diabetes mellitus. Clinical Endocrinology, 2010 Oct; 73(4): 463-8.

Kanazawa I*, Yamaguchi T, Hayashi K, Takase H, Shimizu T, Sugimoto T. Effects of treatment with risedronate and alfacalcidol on progression of atherosclerosis in postmenopausal women with type 2 diabetes mellitus accompanied with osteoporosis. The American Journal of the Medical Sciences, 2010 Jun; 339(6): 519-24.

Kanazawa I*, Yamaguchi T, Yano S, Yamauchi M, Sugimoto T. Fasudil hydrochloride induces osteoblastic differentiation of stromal cell lines, C3H10T1/2 and ST2, via bone morphogenetic protein-2 expression. Endocrine Journal, 2010 May; 57(5): 415-21.

Kanazawa I*, Yamaguchi T, Yamamoto M, Sugimoto T. Relationships between treatments with insulin and oral hypoglycemic agents versus the presence of vertebral fractures in type 2 diabetes mellitus. Journal of Bone and Mineral Metabolism, 2010 Sep; 28(5): 554-60.

Kanazawa I*, Yamaguchi T, Yano S, Yamamoto M, Yamauchi M, Kurioka S, Sugimoto T. Baseline atherosclerosis parameter could assess the risk of bone loss during pioglitazone treatment in type 2 diabetes mellitus. Osteoporosis International, 2010 Dec; 21(12): 2013-8.

Kanazawa I*, Yamaguchi T, Sugimoto T. Baseline serum total adiponectin level is positively associated with changes in bone mineral density after 1 year treatment of type 2 diabetes. Metabolism Clinical and Experimental, 2010 Sep; 59(9): 1252-6.

Kanazawa I*, Yamaguchi T, Yamauchi M, Sugimoto T. Rosuvastatin increased serum osteocalcin levels independent of its serum cholesterol- lowering effect in patients with type 2 diabetes and hypercholesterolemia. Internal Medicine, 2009 Nov; 48(21): 1869-73.

Takase H, Yano S, Yamaguchi T, Kanazawa I, Hayashi K, Yamamoto M, Yamauchi M, Sugimoto T. Parathyroid hormone up-regulates BMP-2 mRNA expression through mevalonate kinase and Rho kinase inhibition in osteoblastic MC3T3-E1 cells. Hormone and Metabolic Research, 2009 Dec; 41(12): 861-5.

Kanazawa I, Yamaguchi T, Yamamoto M, Yamauchi M, Yano S, Sugimoto T. Serum osteocalcin /bone specific alkaline phosphatase ratio is a predictor for the presence of vertebral fractures in men with type 2 diabetes. Calcified Tissue International, 2009 Sep; 85(3): 228-34.

Kanazawa I*, Yamauchi M, Yano S, Imanishi Y, Kitazawa R, Nariai Y, Araki A, Kobayashi K, Inaba M, Maruyama R, Yamaguchi T, Sugimoto T. Osteosarcoma in a pregnant patient with McCune-Albright syndrome. Bone, 2009 Sep; 45(3): 603-8.

Yamaguchi T, Kanazawa I, Yamamoto M, Kurioka S, Yamauchi M, Yano S, Sugimoto T. Associations between components of the metabolic syndrome versus bone mineral density and vertebral fractures in patients with type 2 diabetes. Bone, 2009 Aug; 45(2): 174-9.

Kanazawa I, Yamaguchi T, Yamauchi M, Yamamoto M, Kurioka S, Yano S, Sugimoto T. Adiponectin is associated with changes in bone markers during glycemic control in type 2 diabetes mellitus. The Journal of Clinical Endocrinology and Metabolism, 2009 Aug; 94(8): 3031-7.

Kanazawa I, Yamaguchi T, Yano S, Hayashi K, Yamauchi M, Sugimoto T. Inhibition of the mevalonate pathway rescues the dexamethasone-induced suppression of the mineralization in osteoblasts via enhancing bone morphogenetic protein-2 signal. Hormone and Metabolic Research, 2009 Aug; 41(8): 612-6.

Hayashi K, Yamaguchi T, Yano S, Kanazawa I, Yamauchi M, Yamamoto M, Sugimoto T. BMP/Wnt antagonists are upregulated by dexamethasone in osteoblasts and reversed by alendronate and PTH: potential therapeutic targets for glucocorticoid-induced osteoporosis. Biochemical and Biophysical Research Communications, 2009 Feb; 379(2): 261-6.

Kanazawa I, Yamaguchi T, Yano S, Yamauchi M, Sugimoto T. Activation of AMP-kinase and inhibition of Rho-kinase induce the mineralization of osteoblastic MC3T3-E1 cells through endothelial NOS and BMP-2 expression. American Journal of Physiology Endocrinology and Metabolism, 2009 Jan; 296(1): E139-46.

Kanazawa I, Yamaguchi T, Yamamoto M, Yamauchi M, Yano S, Sugimoto T. Relationships between serum adiponectin levels versus bone mineral density, bone metabolic markers, and vertebral fractures in type 2 diabetes mellitus. European Journal of Endocrinology, 2009 Feb; 160(2): 265-73.

Kanazawa I, Yamaguchi T, Yamamoto M, Yamauchi M, Kurioka S, Yano S, Sugimoto T. Serum osteocalcin level is associated with glucose metabolism and atherosclerosis parameters in type 2 diabetes mellitus. The Journal of Clinical Endocrinology and Metabolism, 2009 Jan; 94(1): 45-9.

Kanazawa I*, Yano S, Takase H, Yamane Y, Yamaguchi T, Sugimoto T. A case of membranous nephropathy associated with chronic sinusitis. Journal of Nephrology, 2009 Mar-Apr; 22(2): 289-94.

In: Proteins Researcher Biographical Sketches … ISBN: 978-1-62100-777-7
Editors: H. Z. Wang and M. Tian

Chapter 26

BIOGRAPHICAL SKETCH

NAME:	TITLE:	DATE OF BIRTH:
Nicola King	**DR.** School of Science and Technology, University of New England, Armidale, NSW 2350. Australia; Bristol Heart Institute, University of Bristol, Bristol Royal Infirmary, Bristol BS2 8HW. UK.	09/03/1970

EDUCATION:

Institution and Location	Degree	Year Conferred	Scientific Field
Napier University, Edinburgh, UK	BSc Hons		
University of Newcastle Upon Tyne, UK	PhD		

CONTACT POINTS:

Address: School of Science and Technology, University of New England, Armidale, NSW 2350

RESEARCH AND PROFESSIONAL EXPERIENCE:

20 year's experience investigating membrane transport in the heart and kidney. More recent research focused on the role of amino acids in the normal and hypertrophic heart:

transport, expression and protection. The emphasis is placed upon expression and activity of amino acid transporters; role of amino acids as myocardial antioxidants and protection against ischaemia reperfusion injury

Professional Appointments:

Lecturer in Biomedical Science (University of New England), member of the British Society for Cardiovascular Research

Honors:

Professional of the year 2011 in Cardiovascular Physiology ("Strathmore's Who's Who" and "America's Registry of Outstanding Professionals"

Publications during Last Three Years:

Books - Editor: *RT-PCR protocols: 2nd Edition* for **Methods in Molecular Biology**. (2010). Humana Press, New Jersey, USA. ISBN: 978-1-60761-628-3.

Peer Reviewed Full Publications:

1. King N., Lin H., & Suleiman M. – S. (2011). Oxidative stress stimulates SNAT1 expression and stimulates cysteine uptake in freshly isolated cardiomyocytes. *Amino Acids.* **40**: 517-526.
2. King N., Lin H., & Suleiman M. – S. (2010). Cysteine protects freshly isolated cardiomyocytes against oxidative stress by stimulating glutathione peroxidase. *Mol Cell Biochem.* **343**: 125-132.
3. King N. (2010). The use of comparative quantitative RT-PCR to investigate the effects of cysteine incubation on GPx1 expression in freshly isolated cardiomyocytes. *Met Mol Biol.* **630**: 215-232.

In: Proteins Researcher Biographical Sketches …
Editors: H. Z. Wang and M. Tian

ISBN: 978-1-62100-777-7

Chapter 27

BIOGRAPHICAL SKETCH

NAME:	TITLE:	DATE OF BIRTH:
Ekaterina V. Konstantinova	RUSSIAN STATE MEDICAL UNIVERSITY	04 Jan. 1972

EDUCATION:

Institution and Location	Degree	Year Conferred	Scientific Field
Russian State Medical University, Moscow		1989-1995	
Internship and Cardiology Residency, Russian Cardiology Research Complex, Moscow		1995-1997,	
Research Fellowship, Russian Cardiology Research Complex		1997-2000	

CONTACT POINTS:

Address: 109-1-45 Leninsky Prosp., 119421 Moscow

RESEARCH AND PROFESSIONAL EXPERIENCE:

1996, 3-month training, electrophysiology laboratory, Baylor College of Medicine, Houston, TX, USA, supervised by Professor Denis Zho,

Researcher. Author of 50 publications. Areas of scientific interest: atherothrombosis, myocardial and cerebral ischemia, inflammation, immune and autoimmune response

In: Proteins Researcher Biographical Sketches … ISBN: 978-1-62100-777-7
Editors: H. Z. Wang and M. Tian

Chapter 28

BIOGRAPHICAL SKETCH

NAME:	TITLE:	DATE OF BIRTH:
Jill M. Lahti	ST. JUDE CHILDREN'S RESEARCH HOSPITAL	1-18-1956

EDUCATION:

Institution and Location	Degree	Year Conferred	Scientific Field
Kalamazoo College	BA	1978	
Baylor College of Medicine	Ph.D.	1986	

CONTACT POINTS:

Address: Department of Tumor Cell Biology 262 Danny Thomas Place MS350, St. Jude Children's Research Hospital, Memphis, TN 38105

Professional Appointments:

1980-86 Graduate Student, Department of Cell Biology, Baylor College of Medicine, Houston, TX

1986-88 Postdoctoral Fellow, Division of Developmental and Clinical Immunology, University of Alabama at Birmingham, AL

1988-90 Research Associate, Division of Developmental and Clinical Immunology, Howard Hughes Medical Institute at University of Alabama at Birmingham, AL

1990-91 Postdoctoral Fellow, Department of Biochemistry, University of Alabama at BirmingÂham, AL

1991-94 Research Associate, Department of Tumor Cell Biology, St. Jude Children's ReÂsearch Hospital, Memphis, TN

1994 -99 Assistant Member, Department of Tumor Cell Biology, St. Jude Children's ReÂ-search Hospital, Memphis, TN

1998 -07 Director, Cancer Center Cell Microinjection and Live Cell Imaging Shared Resource, St. Jude Childrenâ€™s Research Hospital, Memphis, TN

2004-present Associate Professor, Department of Molecular Sciences, College of Medicine, University of Tennessee Health Science

2004-present Director, Cancer Center Cytogenetics Shared Resource, St. Jude Children's Research Hospital, Memphis, TN

1999-present Associate Member, Department of Tumor Cell Biology, St. Jude Childrenâ€™s Research Hospital, Memphis, TN (primary appointment)

Honors:

Service on Review Committees

US Department of Veteran Affairs, Medical Research Service (1999)

Ad hoc Reviewer NIH CDF5 Study Section (Oct 1999)

Wellcome Trust (2000)

Canadian Institute of Health and Research (2002, 2004)

National Science Foundation (2005)

Advanced Research Assessment Program – State of Texas Biological Science II (2006)

Ad hoc Reviewer NIH ZRG1 Onc-L Cancer Diagnostics and Treatment SBIR/STTR. (June 2007)

Ad hoc Reviewer NIH ZRG1 Onc-L Cancer Diagnostics and Treatment SBIR/STTR, (Feb 2007)

Advanced Research Assessment Program – State of Texas on site reviewer for UT Southweestern Metroplex Medical Imaging Center and Innovations in Medical Technology special legislative initiative review (Sept 28-29, 2008)

Publications during Last Three Years:

Guo Q, Xia B, Moshiah S, Xu C, Jiang Y, Chen Y, Sun Y, Lahti JM, Zhang XA. The micro-environmental determinants for epithelial cyst morphogenesis. Eur J Cell Biol, 87: 251-266, 2008.

Loyer P, Trembley JH, Grenet J, Busson A, Corlu A, Kidd VJ, Lahti JM. Characterization of cyclin L1 and L2 interactions with CDK11 and splicing factors: Influence of cyclin L isoforms on splice site selection. J Biol Chem, 283:7721-7732, 2008.

Jiang M, Zhu K, Grenet J, Lahti JM. Activation of Caspase-8 by retinoic acid sensitizes neuroblastoma cells to TNFÎ±- and drug- induced apoptosis. BBA-Mol Cell Res 1783:1055-1067, 2008.

Inoue A, Hyle J, Lechner MS, Lahti JM. Perturbation of HP1 localization and chromatin binding ability causes defects in sister-chromatid cohesion. Mutat Res 657:48-55, 2008.

Sivakolundu SG, Nourse A, Moshiach S, Bothner B, Ashley C, Satumbaâ€Œ J, Lahti JM, Kriwacki RW. Intrinsically unstructured domains of Arf and Hdm2 form bi-molecular oligomeric structures in vitro and in vivo. J Mol Biol 384: 240-254, 2008.

Barbero S, Mieglo A, Torres V, Teitz T, Shields DJ, Mikolon D, Bogyo M, Barila D, Lahti JM, Schlaepfer D, Stupack DG. Caspase-8 association with focal adhesion complex promotes tumor cell migration and metastasis. Cancer Res 69:3755-3763, 2009.

Malumbres M, Harlow E, Hunt T, Hunter T, Lahti JM, Manning G, Morgan DO, Tsai L-H, Wolgemuth DJ. Cyclin-dependent kinases: a family portrait. Nature Cell Biol 11:1275-1276, 2009.

Lagisetti C, Pourpak A, Goronga T, Jian Q, Cui X, Hyle J, Lahti JM, Morris SW, Webb TR. Synthetic mRNA Splicing Modulatory Compounds with In vivo Anti-tumor activity. J Medicinal Chem, 52:6979-6990, 2009.

George RE, Lahti JM, Zhu K, Finkelstein D, Ingle AM, Reid JM, Krailo M, Neuberg D, Adamson PC, Blaney SM, Diller L. Phase 1 Study of Decitabine in Combination with Doxorubicin and Cyclophosphamide in the treatment of relapsed or Refractory Solid Tumors- a Childrenâ€™s Oncology Group Study. Pediatr Blood Cancer. 55:629-638, 2010 PMID: 2058965

Loyer P, Busson A, Trembley JH, Hyle J, Grenet J, Zhao W, Ribault C, Montier T, Kidd VJ, Lahti JM. The RNA binding motif protein 15B (RBM15B/OTT3) is a functional competitor of serine- arginine (SR) proteins and antagonizes the positive effect of the CDK11p110-cyclin L2Î± complex on splicing. J Biol Chem. 286:147-159, 2011. PMID: 21044963

Jiang M, Stanke J, Lahti JM. The connections between neural crest development and neuroblastoma. Curr Top Dev Biol. 94:77-127, 2011. PMID: 21295685

McGann, PT Howard TA, MS[1], Flanagan JM, Lahti JM, Ware RE. Chromosome Damage and Repair in Children with Sickle Cell Anemia and Long-Term Hydroxyurea Exposure. Br. J Haematol. In press, 2011.

Teitz T, Stanke J, Federico S, Bradley CL, Brennan R., Zhang J, Johnson; MD, Sedlacik J, InoueM. Zhang ZM, Frase S, Rehg JE, Hillenbrand CM, Finkelstein D, Calabrese C, Dyer MD, Lahti JM. Preclinical Models for Neuroblastoma: Establishing a Baseline for Treatment . PLoS ONE in press, 2011.

In: Proteins Researcher Biographical Sketches …
Editors: H. Z. Wang and M. Tian
ISBN: 978-1-62100-777-7

Chapter 29

BIOGRAPHICAL SKETCH

NAME:	TITLE:	DATE OF BIRTH:
Stanley S. Levinson	UNIVERSITY OF LOUISVILLE AND VETERAN ADMINISTRATION MEDICAL CENTER	01/20/1939

EDUCATION:

Institution and Location	Degree	Year Conferred	Scientific Field
	Ph. D. UCLA		

Professional Appointments:

Professor of Pathology and Laboratory Medicine

Publications during last three years

1. Levinson SS. Clinical Usefulness of Biomarkers for Predicting Risk. Advances in Clin Chem 2009;48:1-25.
2. Levinson SS, Shaheen SP. Serum free light chain analysis may miss monoclonal light chains that urine immunofixation electrophoreses would detect. In revision Clin Chim Acta 2009;406162-66.

3. Levinson SS. Weak associations between prognostic biomarkers and disease in preliminary studies illustrates the breach between statistical significance and diagnostic discrimination. Clin Chim Acta 2010;411:467-73.
4. Levinson SS. Polyclonal Free Light Chain (FLC) of Ig may Interfere with interpretation of monoclonal FLC k/ï€¯l Ratio. Annals Clin Lab Med. 2010;40:3348-353.
5. Levinson SS. Hook Effect with Lambda Free Light Chain in Serum Free Light Chain Assay. Clin Chim Acta 411;1834-6:2010.
6. Levinson SS. Immunoelectrophoresis in Kaplan and Pesceâ€™s Clinical Chemistry: Theory, Analysis, Correlation, Accessory to 5th ed, methods in analysis: Hickman PE and Koerbin, G. St. Louis, MO: Mosby, 2010, 740-52.(Book)
7. Levinson SS. Bence Jones Proteins in Kaplan and Pesceâ€™s Clinical Chemistry: Theory, Analysis, Correlation, , Accessory to 5th ed, methods in analysis 5th ed: Hickman PE and Koerbin, G. St. Louis, MO: Mosby, 2010. 219-231.(Book)
8. Levinson SS. Immunoelectrophoreisi (version 2.0) In Encyclopedia of Life Sciences (ELS). John Wiley and Sons, Ltd, Chichester PO19 8SQ, UK. 2009. Electronic Publication ; (Book)
http://www3.interscience.wiley.com/cgi-bin/home.

Two publications are in press one in Clin Chem dealing with overestimation of free light chains and a second in Clin Chem Lab med dealing with statistical analysis of biomarkers.

In: Proteins Researcher Biographical Sketches … ISBN: 978-1-62100-777-7
Editors: H. Z. Wang and M. Tian

Chapter 30

BIOGRAPHICAL SKETCH

NAME:	TITLE:	DATE OF BIRTH:
Yang Li	AMGEN INC.	

EDUCATION:

Institution and Location	Degree	Year Conferred	Scientific Field
Stanford University	Ph.D.		

CONTACT POINTS:

Address: 1120 Veterans Blvd, South San Francisco, CA 94080

Professional Appointments:

Scientific Director

Publications during last three years

1) Wu, X. and Li, Y.* (2011) Understanding The Structure-Function Relationship Between FGF19 And Its Mitogenic And Metabolic Activities. In Endocrine FGFs and Klothos, edited by Makoto Kuro-o, Landes Bioscience and Springer Science Media.
2) Gupte, J., Yang, L., Wu, X., Weiszmann, J., Hecht, R., Lemon, B., Lindberg, R., Wang, Z., and Li, Y.* (2011) The FGFR D3 Domain Determines Receptor Selectivity For Fibroblast Growth Factor 21. J. Mol. Biol., 408:491.

3) Swaminath, G., Jaeckel, P., Guo, Q., Cardozo, M., Weiszmann, J., Lindberg, R., Wang, Y., Schwandner, R., and Li, Y.* (2011) Mutational Analysis of G-Protein Coupled Receptor - FFA2. Biochem. Biophys. Res. Comm., 405:122.
4) Wu, X., Ge, H., Lemon, B., Vonderfecht, S., Baribault, H., Weiszmann, J., Gupte, J., Walker, N., Lindberg, R., Wang, Z., and Li Y.* (2010) Separating mitogenic and metabolic activities of FGF19. Proc. Natl. Acad. Sci. USA, 107:14158.
5) Swaminath, G., Jaeckel, P., Guo, Q., Cardozo, M., Weiszmann, J., Lindberg, R., Wang, Y., Schwandner, R., and Li, Y.* (2010) Allosteric rescuing of loss-of-function FFA2 mutations. FEBS Lett., 584:4208.
6) Ge, H., Xiong, Y., Lemon, B., Lee, K. J., Tang, J., Wang, P., Weiszmann, J., Hawkins, N., Laudemann, J., Min, X., Penny, D., Wolfe, T., Liu, Q., Zhang, R., Yeh, W.-C., Shen, W., Lindberg, R., Wang, Z., Sheng, J., and Li, Y.* (2010) Generation of novel long-acting globular adiponectin molecules. *J.* Mol. Biol., 399:113.
7) Wu, X., Ge, H., Lemon, B., Vonderfecht, S., Weiszmann, J., Hecht, R., Gupte, J., Hager, T., Wang, Z., Lindberg, R., and Li, Y.* (2010) FGF19 induced hepatocyte proliferation is mediated through FGFR4 activation. J. Biol. Chem., 285:5165.
8) Wang, Y., Jiao, X., Kayser, F., Liu, J., Wang, Z., Wanska, M., Greenberg, J., Weiszmann, J., Ge, H., Tian, H., Wong, S., Schwandner, R., Lee, T., and Li, Y.* (2010) The first synthetic agonists of FFA2: discovery and SAR of phenylacetamides as allosteric modulators. Bioorg Med Chem Lett., 20:493.
9) Wu, X. and Li, Y.* (2009) Role of FGF19 induced FGFR4 activation in the regulation of glucose homeostasis. Aging, 1:1023.
10) Wu, X., Ge, H., Lemon, B., Weiszmann, J., Gupte, J., Hawkins, N., Li, X., Tang, J., Lindberg, R., and Li, Y.* (2009) Selective activation of FGFR4 by a novel FGF19 does not improve glucose metabolism in *ob/ob* mice. Proc. Natl. Acad. Sci. USA, 106:14379.
11) Xu, J., Stanislaus, S., Chinookoswong, N., Lau, Y., Hager, T., Patel, J., Ge, H., Weiszmann, J., Lu, S.-C., Graham, M., Busby, J., Hecht, R., Li, Y.-S., Li, Y., Lindberg, R., and Veniant, M. (2009) Acute glucose-lowering and insulin-sensitizing action of FGF21 in insulin resistant mouse models-association with liver and adipose tissue effects. AJP-Endocrinology and Metabolism, 297:E1105.
12) Li, X., Ge, H., Weiszmann, J., Hecht, R., Li, Y-S., Veniant, M., Xu, J., Wu, X., Lindberg, R., and Li Y.* (2009) Inhibition of Lipolysis May Contribute To The Acute Regulation Of Plasma FFA and Glucose By FGF21 In 013/0b Mice. FEBS Lett, 583:3230.
13) Connors, R.V., Wang, Z., Harrison, M., Zhang, A., Wanska, M., Hiscock, S., Fox, B., Dore, M., Labelle, M., Sudom, A., Johnstone, S., Liu, .1., Walker, N.P., Chai, A., Siegler, K., Li, Y., and Coward, P. (2009) Identification of a PPARy agonist with partial agonistic activity on PPARy. Bioorg Med Chem Lett., 19:3550.
14) Motani, A., Wang, Z., Weiszmann, J., McGee, L.R., Lee, G., Liu, Q., Staunton, J., Fang, Z., Fuentes, H., Lindstrom, M., Liu, J., Biermann, D., Jaen, J., Walker, N., Learned, R.M., Chen, J.-L., and Li, Y.* (2009) INT131: A Selective Modulator of PPARy. J. Mol. Biol., 386:1301.
15) Lee, T., Schwandner, R., Swaminath, G., Weiszmann, J., Cardozo, M., J., Greenberg, J., Jaeckel, P., Ge, H., Wang, Y., Jiao, X., Liu, J., Kayser, F., Tian, H., and

Li, Y.* (2008) Identification and Functional Characterization of Allosteric Agonists for the G Protein Coupled Receptor FFA2. Mol. Pharm., 74:1599.
16) Wu, X., Lemon, B., Li, X., Gupte, J., Weiszmann, J., Stevens, J., Hawkins, N., Shen, W., Lindberg, R., Chen, Hui, T., and Li. Y.* (2008) C-Terminal Tail of FGF19 Determines Its Specificity Toward Klotho Co-receptors. J. Biol. Chem., 283:33304.
17) Ge, H., Li, X., Weiszmann, J., Wang, P., Baribault, H., Chen, 1.-L., Tian, H., Li, Y.* (2008) Activation of GPR43 In Adipocytes Leads to Inhibition of Lipolysis and Suppression of Plasma Free Fatty Acids. Endocrinology, 149:4519.
18) Li, Y.* Wang, Z., Furukawa, ***N.,*** Escaron, P., Weiszmann, J., Lee, G., Lindstrom, M., Liu, J., Liu, X., Xu, H., Plotnikova, 0., Prasad, V., Walker, N., Learned, M., and Chen, J.-L. (2008) T2384, A Novel Antidiabetic Agent with Unique PPARy Binding Properties. J. Biol. Chem., 283:9168. *Also corresponding author.
19) Ge, H., Weiszmann, J., Reagan, J.D., Gupte, J., Baribault, H., Gyuris, T., Chen, J.-L., Tian, H., and Li. Y.* (2008) Elucidation Of Signaling and Functional Activities of An Orphan GPCR-GPR81. J. Lipid Res., 49:797.

In: Proteins Researcher Biographical Sketches ... ISBN: 978-1-62100-777-7
Editors: H. Z. Wang and M. Tian

Chapter 31

BIOGRAPHICAL SKETCH

NAME:	TITLE:	DATE OF BIRTH:
Luísa Lobato	**DEPARTMENT OF NEPHROLOGY AND UNIDADE CLINICA DE PARAMILOIDOSE, HOSPITAL DE SANTO ANTONIO-CENTRO HOSPITALAR DO PORTO, PORTO, PORTUGAL**	**AUGUST 1, 1962**

EDUCATION:

Institution and Location	Degree	Year Conferred	Scientific Field
Institute of Biomedical Sciences Abel Salazar, University of Porto, Porto, Portugal	MD		
Health Sciences at Institute of Biomedical Sciences Abel Salazar, University of Porto, Porto, Portugal.	PhD		

CONTACT POINTS:

Address: Largo Prof. Abel Salazar, 4099-001, Porto, Portugal. E-mail: llobato@sapo.pt phone +351 222 074 684 Fax: +351 222 059 125

RESEARCH AND PROFESSIONAL EXPERIENCE:

Clinical, histology and genetic epidemiology of nephropathic amyloidosis. Genetics of inherited renal diseases: gender susceptibility, familial aggregation and clinical heterogeneity. Dysfunction of renal tubular cells and urinary biomarkers in amyloidosis. Control of erythropoietin production.

Kidney and liver-kidney transplantation. Specific factors contributing to renal graft loss.

Professional Appointments:

2002 - present Graduate Consultant of Nephrology, Hospital de Santo António-Centro Hospitalar do Porto, Porto, Portugal

1995 - 2002 Assistant of Nephrology, Hospital de Santo António-Centro Hospitalar do Porto, Porto, Portugal

1989 - 1995 Fellow of Nephrology, Hospital de Santo António-Centro Hospitalar do Porto, Porto, Portugal

2004 - present Co-ordenation of outpatient clinic of Nephrology of familial amyloidotic polyneuropathy, Hospital de Santo António-Centro Hospitalar do Porto, Porto, Portugal

2000 - present Activity on simultaneous liver-kidney transplantation, Hospital de Santo António-Centro Hospitalar do Porto, Porto, Portugal

1995 -present Activity on kidney transplantation, Hospital de Santo António-Centro Hospitalar do Porto, Porto, Portugal

2000 - present Activity on histopathology and immunohistochemistry of renal disorders, Hospital de Santo António-Centro Hospitalar do Porto, Porto, Portugal

2002 - present Professor of Nephrology, Institute of Biomedical Sciences Abel Salazar, University of Porto, Portugal

1995 - 2002 Assistant Professor of Nephrology, Institute of Biomedical Sciences Abel Salazar, University of Porto, Portugal

1986 - 1990 Clinical Investigator of Medical Genetics, Institute of Biomedical Sciences Abel Salazar, University of Porto, Portugal

Honors:

2011: Best work of the congress in the area of clinical Nephrology, Portuguese Society of Nephrology

2005: Best presentation in nineteenth Portuguese Congress of Nephrology.

2004: Fresenius Prize from the Portuguese Society of Nephrology

2004: Award for Investigation from Hospital de Santo António-Centro Hospitalar do Porto

2002: Roche Prize for Portuguese Neprologists

1999: Best Oral Communication of Clinical Nephrology, XIII Congress of Portuguese Society of Nephrology

1998: Gulbenkian Prize for Prize for young investigators

1997: European Renal Association Grant

Publications during Last Three Years:

Rocha A, Lobato L, Silva H, Beirão I, Santos J, Pessegueiro H, Almeida R, Cabrita A. Characterization of end-stage renal disease after liver transplantation in transthyretin amyloidosis (ATTR V30M). Transplant Proc. 2011 Jan-Feb;43(1):189-93.

Fonseca I, Almeida M, Martins LS, Santos J, Dias L, Lobato L, Henriques AC, Mendonça D. First-year renal function predicts long-term renal allograft loss. Transplant Proc. 2011 Jan-Feb;43(1):106-12.

Beirão I, Moreira L, Barandela T, Lobato L, Silva P, Gouveia CM, Carneiro F, Fonseca I, Porto G, Pinho E Costa P. Erythropoietin production by distal nephron in normal and familial amyloidotic adult human kidneys. Clin Nephrol. 2010 Nov;74(5):327-35.

Beirão I, Lobato L, Moreira L, Mp Costa P, Fonseca I, Cabrita A, Porto G. Long-term treatment of anemia with recombinant human erythropoietin in familial amyloidosis TTR V30M. Amyloid. 2008 Sep;15(3):205-9.

Beirão I, Moreira L, Porto G, Lobato L, Fonseca I, Cabrita A, Costa PM. Low erythropoietin production in familial amyloidosis TTR V30M is not related with renal congophilic amyloid deposition. A clinicopathologic study of twelve cases. Nephron Clin Pract. 2008;109(2):c95-9

ISBN: 978-1-62100-777-7

Chapter 32

BIOGRAPHICAL SKETCH

NAME:	TITLE:	DATE OF BIRTH:
Angelo Lupo	**1) DIPARTIMENTO DI SCIENZE BIOLOGICHE ED AMBIENTALI, FACOLTA DI SCIENZE, UNIVERSITA DEL SANNIO, VIA PORT?ARSA 11, 82100 BENEVENTO, ITALY 2) DIPARTIMENTO DI BIOCHIMICA E BIOTECNOLOGIE MEDICHE UNIVERSITA FEDERICO II, VIA S. PANSINI 5, 80131 NAPOLI, ITALY**	**SEPTEMBER 22TH, 1956**

EDUCATION:

Institution and Location	Degree	Year Conferred	Scientific Field
	University degree		Biological Science

CONTACT POINTS:

Address: Dipartimento di Scienze Biologiche ed Ambientali, Facoltà di Scienze, Università del Sannio, via Port?Arsa 11, 82100 Benevento, Italy

RESEARCH AND PROFESSIONAL EXPERIENCE:

Since 1981 until 1988 Angelo Lupo was recipient of a post-degree fellowship from the National Research Council(C.N.R.), Rome, Italy. In 1988 he was recipient of a Postdoctoral Fellowship at the European Molecular Biology Laboratory,Heidelberg (Germany) within" Gene Structure and Regulation Programme ". Since 2001 Angelo Lupo is Associate Professor of Biochemistry at Department of Biological Sciences, Università del Sannio ? Benevento.

Angelo Lupo developed his scientific interests in the field of the transcriptional control in the regulation of the gene expression in Eukaryotes. Particularly, he carried out four main research projects:

1) Transcriptional regulation of human Aldolase A gene.
2) Structural and functional characterization of the transcriptional repressor ZNF224, a modulator of human Aldolase a gene transcription.
3) Transcriptional control of Retinol-Binding Protein expression gene
4) Identification and biological characterization of natural compounds from plant or vegetables showing antiproliferative and apoptotic properties.

Angelo Lupo is coauthor of 17 scientific papers published on international journals with peer review.

Publications during Last Three Years:

1) Bianconcini, A, Lupo A., Capone, S, Quadro, L, Monti, M, Zurlo, D, Fucci, A, Sabatino, L, Brunetti, A, Chiefari, E, Gottesman, M. E, Blaner, W.S, and Colantuoni, V. (2009). Transcriptional activity of the murine retinol binding protein gene is regulated by a multiprotein complex containing HMGA1, p54nrb/NonO, protein -associated splicing factor (PSF) and steroidogenic factor 1 (SF1) / liver receptor homologue1(LRH-1). The Int. J. Biochem. Cell Biol., vol. 41; p. 2189-2203
2) Chiefari, E, Paonessa, F, Iritano, S, Le Pera, I, Palmieri, D, Brunetti, G, Lupo A., A, Colantuoni, V, Foti, D, Gulletta, E, De Sarro, G, Fusco, A, And Brunetti, A (2009). The cAMP-HMGA1-RBP4 system: a novel biochemical pathway for modulating glucose homeostasis. BMC Biol., vol. 21; p. 7-24
3) Cesaro E, De Cegli R, Medugno L, Florio F, Grosso M, Lupo A., Izzo P, Costanzo P (2009). The Kruppel-like zinc finger protein ZNF224 recruits the arginine methyltransferase PRMT5 on the transcriptional repres sor complex of the aldolase A gene. J.Biol. Chem., 284, 32321-32330
4) Lupo A, Cesaro E, Montano G, Izzo P, Costanzo P. (2011) ZNF224: Structure and role of a multifunctional KRAB-ZFP protein. Int J Biochem Cell Biol. Doi: 10.1016/J.biocel 2010.12.020

In: Proteins Researcher Biographical Sketches ...
Editors: H. Z. Wang and M. Tian
ISBN: 978-1-62100-777-7

Chapter 33

BIOGRAPHICAL SKETCH

NAME:	TITLE:	DATE OF BIRTH:
Jillian Madine	INSTITUTE OF INTEGRATIVE BIOLOGY, UNIVERSITY OF LIVERPOOL	18/11/80

EDUCATION:

Institution and Location	Degree	Year Conferred	Scientific Field
University of Manchester	PhD		

CONTACT POINTS:

Address: Biosciences Building, Crown Street, Liverpool, L69 7ZB

Publications during Last Three Years:

R. Edwards, J. Madine, L. Fielding and D.A. Middleton, Measurement of multiple torsional angles from one-dimensional solid-state NMR spectra: application to the conformational analysis of a ligand in its biological receptor site, Phys. Chem. Chem. Phys. 2010, 12, 13999-14008

J. Madine, D.A. Middleton, Comparison of aggregation enhancement and inhibition as strategies for reducing the cytotoxicity of the aortic amyloid polypeptide medin, European Biophysics Journal, 2010, 39, 1281-1288

J.Madine, D.A. Middleton, Targeting a-synuclein aggregation for Parkinson's disease treatment, Drugs of the Future, 2009, 34, 655-663 Invited article

J. Madine, X. Wang, D.R. Brown, D.A. Middleton, Evaluation of b-alanine- and GABA-substituted peptides as inhibitors of disease-linked protein aggregation, ChemBioChem, 2009, 10, 1982-1987

J. Madine, A. Copland, L.C. Serpell, D.A. Middleton, Cross-b spine architecture of fibrils formed by the amyloidogenic segment NFGSVQFV of medin from solid-state NMR and X-ray fibre diffraction measurements. Biochemistry, 2009, 48, 30893099

J. Madine, J.C. Clayton, E.A. Yates, D.A. Middleton, Exploiting a 13C-labeled heparin derivative for solid-state NMR investigations of peptide-glycan interactions within amyloid fibrils. Organic & Biomolecular Chemistry, 2009, 7, 2414-2420

H. Amijee, J. Madine, D.A. Middleton, A.J. Doig, Inhibitors of protein aggregation and toxicity. Biochemical Society Transactions, 2009, 37, 692-696

J. Madine, E. Hughes, A.J. Doig, D.A. Middleton, The effects of á-synuclein on phospholipid vesicle integrity: a study using 31P NMR and electron microscopy. Molecular Membrane Biology, 2008, 25, 518527

J. Madine, E. Jack, P. Stockley, S. Radford, L. Serpell, D.A. Middleton, Structural insights into the polymorphism of amyloid-like fibrils formed by region 20-29 of amylin revealed by solid-state NMR and X-ray fibre diffraction. Journal of American Chemical Society, 2008, 1499015001

J. Madine, A.J. Doig, D.A. Middleton, Design of an N-methylated peptide inhibitor of á-synuclein aggregation guided by solid-state NMR. Journal of American Chemical Society, 2008, 130, 7873-7881 Spotlighted in ACS Chemical Biology and SciBX: Science-Business eXchange

In: Proteins Researcher Biographical Sketches ... ISBN: 978-1-62100-777-7
Editors: H. Z. Wang and M. Tian

Chapter 34

BIOGRAPHICAL SKETCH

NAME:	TITLE:	DATE OF BIRTH:
Ana Laura Martinez-Hernandez	INSTITUTO TECNOLOGICO DE QUERETARO	MAY, 10TH 1972

EDUCATION:

Institution and Location	Degree	Year Conferred	Scientific Field
Universidad Autonoma de Queretaro BSc	PhD		Materials Engineering
Chemical Engineering from Instituto Tecnologico de Queretaro	MSc		

CONTACT POINTS:

Address: Av. Tecnologico S/No. esq. Mariano Escobedo, Col. Centro Historico, C.P. 76000, Queretaro, Queretaro, Mexico.

RESEARCH AND PROFESSIONAL EXPERIENCE:

June 2002 September 2002: Research Stay, NU NASA URETI BIMat Center, Department of Mechanical Engineering, Northwestern University, Evanston, Illinois.
August 2003 July 2005: Postdoctoral fellow, Laboratory of Advanced Polymers and Optimized Materials, University of North Texas, Denton, Texas.
From April 2003: Research and Professor, Instituto Tecnologico de Queretaro.
From January 2000: Industrial Consultant

Professional Appointments:

1998 1999: Chemical Process Engineer, Instituto Mexicano del Petroleo and Coorporacion Mexicana de Investigacion en Materiales.

Honors:

First place, poster session, XVIII International Materials Research Congress, with the work entitled: Keratin Modified Multiwalled Carbon Nanotubes: Grafting And Non Covalent Interactions. 2009

Second place, poster session, XVIII International Materials Research Congress, with the work entitled: Novel Membranes Based On Keratin Extracted From Chicken Feathers And Polyurethane To Remove Cr(VI). 2009

Awarded by L'oreal-UNESCO-Mexican Academy of Sciences as National Fellow, For Women in Science, 2007.

Third place in Alejandrina Contest for Technological and Scientific Research, Universidad Autonoma de Queretaro, 2004.

Awarded by American Chemical Society and Mexican Academy of Science with Summer Research Stay Fellow, 2004.

Publications during Last Three Years:

Scientific articles:

A. Espíndola-González, A.L. Martínez-Hernández, V.M. Castaño, C. Angeles-Chávez, C. Velasco-Santos, Novel crystalline SiO2 nanoparticles via annelids bioprocessing of agro-industrial wastes, Nanoscale Research Letters 5, 1408, 2010.

Carlos Velasco-Santos, Ana L. Martínez-Hernández, Witold Brostow, Víctor M. Castaño, Influence of Silanization Treatment on Termo-mechanical properties of multiwalled carbon nanotubes-poly(methylmethacrylate) nanocomposites, Journal of Nanoscience and Nanotechnology, In Press 2011

Ana L. Martínez-Hernández, Carlos Velasco-Santos, Víctor M. Castaño, Carbon Nanotubes Composites: Processing, Grafting and Mechanical and Thermal Properties, Current Nanoscience, 6, 12, 2010.

V. Saucedo-Rivalcoba, A.L. Martínez-Hernández, G. Martínez-Barrera, C. Velasco-Santos, V.M. Castaño, Chicken feathers. Applied Physics A. Published on line December 2010.

V. Saucedo-Rivalcoba, A.L. Martínez-Hernández, G. Martínez-Barrera, C. Velasco-Santos, V.M. Castaño, J.L. Rivera-Armenta Removal of Hexavalent Chromium from water by polyurethane-keratin hybrid membranas. Water, air and soil pollution, Published on line November 2010.

Gonzalo Martínez-Barrera, Ana Laura Martínez-Hernandez, Carlos Velasco-Santos, Witold Brostow, Polymer concretes improved by fiber reinforced and gamma irradiation, e-Polymers, no. 103, August 2009.

Gaurav Mago, Carlos Velasco-Santos, Ana L. Martínez-Hernández, Dilhan M. Kalyon, Frank T. Fisher, Effect of Functionalization on the Crystalline Behavior of MWNT-PBT Nanocomposites, Materials Research Society Symposium Proceedings, 1056-HH11-35, 2008.

G. de la Rosa, H. E. Reynel-Avila, A. Bonilla-Petriciolet, I. Cano-Rodríguez, C. Velasco-Santos, A.L. Martínez-Hernandez, Recycling Poultry Feathers for Pb Renoval from Wastewater: Kinetic and Equilibrium Studies, International Journal of Chemical and Biomolecular Engineering, 1, 185, 2008

A.L. Martínez-Hernández, A.L. Santiago-Valtierra, M.J. Alvarez-Ponce, Chemical modification of keratin biofibers by graft polymerization of methyl methacrylate using redox initiation, Materials Research Innovations, 12, 184, 2008, Factor de Impacto 0.89

Chapter books:

Velasco-Santos C., Martínez-Hernández A.L., Castaño V.M., (ed.) James N. Ling, Carbon Nanotube-Polymer Nanocomposites: Principles and Applications, Capítulo por invitación. En "Nanotechnology Research Collection". Publicado por Nova Science Publishers Inc. ISBN: 978-1-60741-292-2 , 2009.

Martínez-Hernández A.L., Velasco-Santos C., Castaño V.M. (Eds.) V.A. Basiuk and E.V. Basiuk,. Advanced carbon nanotube-based nanocomposites: Principles, synthesis and chemical-modification. Capítulo por Invitación. En "Chemistry of Carbon Nanotubes" American Scientific Publishers. ISBN:1-58883-128-0, 2008.

In: Proteins Researcher Biographical Sketches ...
Editors: H. Z. Wang and M. Tian
ISBN: 978-1-62100-777-7

Chapter 35

BIOGRAPHICAL SKETCH

NAME:	TITLE:	DATE OF BIRTH:
Tara McMorrow	**DR.** **UNIVERSITY COLLEGE DUBLIN**	**13/01/1971**

EDUCATION:

Institution and Location	Degree	Year Conferred	Scientific Field
University College Galway, Ireland	BSc		
University College Galway, Ireland	PhD		

CONTACT POINTS:

Address:
Renal Disease Research Group
UCD School of Biomolecular and Biomedical Science,
UCD Conway Institute,
University College Dublin,
Belfield, Dublin 4
Ireland

RESEARCH AND PROFESSIONAL EXPERIENCE:

2002- to date: College Lecturer in <?xml:namespace prefix = st1 ns = "urn:schemas-microsoft-com:office:smarttags" />UCD School of Biomolecular and Biomedical Science and Conway Investigator, Conway Institute, University College Dublin, Ireland
1998-2002: Senior Research Fellow, University College Dublin, Ireland

1995-1998: EU Research Scientist, Dept. Cell Biology, Erasmus University, Rotterdam, Holland

Activities:

Principle investigator and Co-investigator on numerous Irish and EU funded grants
Reviewer for Diabetes Trust UK
Reviewer for UCD Internal Funding Scheme
Member of organising committee for UCD Access Science annual competition
Co-organiser of European Renal Cell Study Group, Dublin, Ireland, March 2006
Co-organiser of PREDICTOMICS FP6 Annual Meeting, Dublin, Ireland, May 2005
Co-organiser of CARCINOGENOMICS FP6 Annual Meeting, Dublin, Ireland, Dec 2008
Member of Medical Degree Programme Systems II Board: 2005 to date
Member of Ireland ECOPA Platform Development Committee: 2005 to present
Reviewer for several scientific journals including Journal of American Society of Nephrology; Kidney International; Nephrology, Dialysis, Transplantation; Nephron; American Journal of Physiology; FEBs Letters

Professional Appointments:

Irish Nephrological Society member
Irish Research Scientists Association member
Irish Society of Toxicology board member
American Society of Nephrology member

Honors:

EU SysKid Co-Investigator Award 2010
Irish Government IRCSET Award 2009
UCD Seed Funding PI Award 2008
EU Carcinogenomics Co-Investigator Award 2006
Science Foundation Ireland Conference Award 2006
Science Foundation Ireland Research Project PI Award 2004
Health Research Board Basic Reseach Grant PI Award 2003
Irish Nephrological Society Amgen Research Bursary award, 2003
HEA Dublin Molecular Medicine Centre Fellowship award, 2002
HEA Conway Institute Research Fellowship Award, 2000
EU Post-Doctoral Research Fellowship Award,1995-1998
Enterprise Ireland Postgraduate Fellowship Award from 1991-1994

Publications during Last Three Years:

O'Connell S, Slattery C, Ryan MP, McMorrow T. (2011) Identification of novel indicators of cyclosporine A nephrotoxicity in a CD-1 mouse model. Toxicol Appl Pharmacol. 252(2):201-10.

Ellis JK, Athersuch TJ, Cavill R, Radford R, Slattery C, Jennings P, McMorrow T, Ryan MP, Ebbels TM, Keun HC. (2011). Metabolic response to low-level toxicant exposure in a novel renal tubule epithelial cell system. *Mol Biosyst.* 7(1):247-57.

Lynch J, Nolan S, Slattery C, Feighery R, Ryan MP, McMorrow T. (2010). High-mobility group box protein 1: a novel mediator of inflammatory-induced renal epithelial-mesenchymal transition. *Am J Nephrol.* 32(6):590-602.

N. Martin-Martin, G. Ryan, T. McMorrow, M.P. Ryan MP. (2010) Sirolimus and cyclosporine A alter barrier function in renal proximal tubular cells through stimulation of ERK1/2 signaling and claudin-1 expression. *Am J Physiol Renal Physiol.* 298(3):F672-82.

P. Jennings, S. Aydin , J. Bennett, R. McBride, C. Weiland, N. Tuite, L.N. Gruber, P. Perco, P.O. Gaora, H. Ellinger-Ziegelbauer, H.J. Ahr, C.V. Kooten, M.R. Daha, P. Prieto, M.P. Ryan, W. Pfaller , T. McMorrow. (2009) Inter-laboratory comparison of human renal proximal tubule (HK-2) transcriptome alterations due to Cyclosporine A exposure and medium exhaustion. Toxicol. In Vitro. 23(3):486-99.

A. Oâ€™ Riordan, T. McMorrow, O. Johnston, W. Gallagher, P. Maguire, G. Cagney, J. Hegarty, A. McCormick, A.J. Watson and M.P. Ryan. (2008) Using SELDI-TOF-MS Apo A1 is a serum biomarker of chronic kidney disease, post orthotopic liver transplantation. *Proteomics: Clinical Applications* 2(9): 1191-1202.

R.Feighery, P. Maguire, M.P. Ryan and T. McMorrow. (2008) Proteomic approach to immune-mediated epithelial-mesenchymal transition (EMT). *Proteomics: Clinical Applications* 2(7): 1110-1117.

L. Nee, S. Oâ€™Connell, S. Nolan, M.P. Ryan and T. McMorrow. (2008) Nitric oxide involvement in TNF-alpha and IL-1 beta-mediated changes in human mesangial cell MMP-9 and TIMP-1. *Nephron: Experimental Nephrology* 110(2):59-66.

C. Slattery, M.P. Ryan and T. McMorrow. (2008) E2A proteins - regulators of cell phenotype in normal physiology and disease. *International Journal of Biochemistry & Cell Biology* 40(8):1431-6.

C. Slattery, M.P. Ryan and T. McMorrow. (2008) Protein kinase C beta overexpression in human renal proximal tubular cells: isoform specific effects and potential roles in renal fibrosis. *International Journal of Biochemistry & Cell Biology* 40(10):2218-29.

In: Proteins Researcher Biographical Sketches ...
Editors: H. Z. Wang and M. Tian
ISBN: 978-1-62100-777-7

Chapter 36

BIOGRAPHICAL SKETCH

NAME:	TITLE:	DATE OF BIRTH:
Bruce Daniel Murphy	**PROFESSOR AND DIRECTOR**	

EDUCATION:

Institution and Location	Degree	Year Conferred	Scientific Field
Colorado State University	BSc	June 1965	Biology
Colorado State University	MSc	August 1969	Physiology
University of Saskatchewan	PhD	March 1973	Reproductive Biology

CONTACT POINTS:

Address:
Université de Montréal
Biomédecine vétérinaire / Centre de recherche en reproduction animale (CRRA)
3200 Sicotte,
St-Hyacinthe, Qc
J2S 7C6

Professional Appointments:

1991 : Director, Centre de recherche en reproduction animale, Faculté de médecine vétérinaire, Université de Montréal.

1991 : Professor with tenure, Département de biomédecine vétérinaire, Faculté de médecine vétérinaire, Université de Montréal.

2005 : Adjunct Professor, Département d'obstétrique-gynécologie. Faculté de médecine, Université de Montréal

2000 : Visiting Scientist, Institut de génétique et de biologie moléculaire et cellulaire, Université Louis Pasteur, Strasbourg, France

1987-91: Founder and Director, Reproductive Biology Research Unit, College of Medicine, University of Saskatchewan,

1987-1991: Professor, Department of Obstetrics and Gynecology, College of Medicine, University of Saskatchewan,

1980: Visiting Associate Professor, Department of Physiology, New York State College of Veterinary Medicine, Cornell University, Ithaca, NY

1982-1987: Professor of Biology, University of Saskatchewan

1977-1982: Associate Professor of Biology, University of Saskatchewan

1973-1977: Assistant Professor of Biology, University of Saskatchewan

1972-1973: Assistant Professor of Anatomy, WAMI Medical Program, University of Idaho

Awards and Honors

2009-2013: Editor-in-Chief, *Biology of Reproduction,* 2009-2013

2010: Award for Excellence in Reproductive Medicine, Canadian Fertility and Andrology Society

2010: Career Achievement Award presented at the III International Symposium on Animal Biology of Reproduction 2010

2009 : *Quebec Science*, Top Ten Discoveries

2009 : Laureate, Fonds québécoise de la recherche sur la nature et les technologies (FQRNT),

2009: Billie A. Field Memorial Lectureship, Univ. Illinois

2007: Distinguished Service Award, Society for the Study of Reproduction,

2006: Elected Fellow, Canadian Academy of Health Sciences,

2004: Pfizer Award for Research Excellence ,

2003: Distinguished Scientist Lecture, College of Medicine, Univ. S. Ala.,

1996: Pfizer Award for Research Excellence,

1990: Wyeth Award, (with P.J. Chedrese et al.),

1988: Elected Fellow, Argentine Academy of Agricultural and Veterinary Science,

1979: Research Associate Award, International Development Research Centre,

1970: Musk Ox Scholarship, Institute for Northern Studies,

1969: University of Saskatchewan Graduate Studies Scholarship,

1969: Hill Foundation Fellowship,

1960: National Merit Society USA,

Publications Last 3 Years:

Total of 22 scientific publications for the last 3 years.

2011

1. Lefèvre, P.L.C., Palin, M-F., and Murphy B.D. 2011. Polyamines on the reproductive landscape. Endocrine Reviews (accepted)
2. Lefèvre, P.L.C., Palin, M-F., Chen, G., Turecki, G., and Murphy, B.D. 2011. Polyamines are implicated in the emergence of the embryo from obligate embryonic diapause. Endocrinology 152:1627-1639.
3. Lefèvre, P.L.C., Palin, M-F., Beaudry, D., Dobias-Goff, M., Desmarais, J.A., Llerena-Vargas, E., and Murphy, B.D. 2011. Uterine signaling at the emergence of the embryo from obligate diapause. Am. J. Physiol. 300:E800-808.
4. Verduzco, A., Fecteau, G., Lefebvre, R., Smith, L.C. and Murphy, B.D. 2011. Expression of steroidogenic proteins in bovine placenta during the first third of gestation. Rep. Fert. Dev. (in press)

2010

5. Murphy, B.D. 2010. Revisiting Reproduction: What a difference a gene makes. Nature Medicine 16:527-529.
6. Bertolin, K., Bellefleur, A-M., Zhang, C., and Murphy, B.D. 2010. Orphan nuclear receptor regulation of reproduction. Anim. Reprod. 7:146-153.
7. Miranda Jiménez, L, Binelli, M., Bertolin, K., Pelletier, R.M. and Murphy. B.D. 2010. Scavenger receptor B-1 and luteal function in mice. J. Lipid Res. 51:2362-2371.
8. Binelli, M., and Murphy, B.D. 2010. Coordinated regulation of follicle development by germ and somatic cells. Rep. Fert. Dev. 22:1-12.
9. Brochu-Goudreau, K., Rehfeldt, C., Blouin, R., Bordignon V., Murphy, B.D., and Palin, M-F. 2010. Adiponectin action from head to toe. Endocrine 37:11-32.
10. Houde, A.A., Méthot, S., Murphy, B.D., Bordignon, V. and Palin, M.F. 2010. Relationships between backfat thickness and reproductive efficiency of sows: a two year trial on two commercial herds fixing their backfat thickness at breeding. *Can. J. Anim. Sci.* 90(3):429-436.

2009

11. Labrecque, B., Beaudry, D., Mayhue, M., Hallé, C., Bordignon, V., Murphy, B.D. and Palin, M.F. 2009. Molecular characterization and expression analysis of the porcine paraoxonase 3 (PON3) gene. Gene. 443:110-120.
12. Labrecque, B., Mathieu, O., Bordignon, V., Murphy, B.D. and Palin, M.F. 2009. Identification of differentially expressed genes in a porcine *in vivo* model of adipogenesis using suppression subtractive hybridization. Comp Biochem Physiol Part D Genomics Proteomics. 4(1):32-44.
13. Duggavathi, R., and Murphy, B.D. 2009. Development. Ovulation Signals. Science. 324:890-91.
14. Lefèvre P.L. and Murphy, B.D. 2009. Differential gene expression in the uterus and blastocyst during the reactivation of embryo development in a model of delayed implantation. Methods Mol. Biol. 550:11-16.

2008

15. Seneda, M.M., Godmann, M., Murphy, B.D., Kimmins, S. and Bordignon, V. 2008 Developmental regulation of histone H3 methylation at lysine 4 in the porcine ovary. Reproduction 135:829-838.

16. Campos, D.B., Palin, M-F., Bordignon, V., and Murphy, B.D. 2008. The beneficial adipokines in reproduction and fertility. Int. J. Obesity 32:223-231.
17. Gévry, N., Schoonjans, K., Guay F., and Murphy, B.D. 2008. Cholesterol supply and sterol regulatory element binding proteins modulate transcription of the Niemann-Pick C1 gene in steroidogenic tissues. J. Lipid Res. 49: 1024-1033.
18. Garver, W.S., Jelenik, D., Francis, G.A., and Murphy, B.D. 2008. The Niemann-Pick C1 gene is downregulated by feedback inhibition of the SREBP pathway in human fibroblasts. J. Lipid Res. 49:1090-1102.
19. Duggavathi, R., Volle, D.H., Mataki, C. Antal, M.C. Messadeq, N., Auwerx, J., Murphy, B.D. and Schoonjans, K. 2008. The nuclear receptor liver receptor homolog-1 is essential for ovulation. Genes and Development. 22: 1871-1876.
20. Houde, A.A., Murphy, B.D., Mathieu, O., Bordignon, V. and Palin, M.F. 2008. Characterization of swine adiponectin and adiponectin receptor polymorphisms and their association with reproductive traits. *Animal Genetics* 39(3):249-57.
21. Palin, M.F., Labrecque, B., Beaudry, D., Mayhue, M., Bordignon, V. and Murphy, B.D. 2008. Visfatin expression is not associated with adipose tissue abundance in the porcine model. *Domestic Animal Endocrinology* 35(1):58-73.
22. Chappaz, E., Albornoz, M.S., Campos, D., Che, L., Palin, M.F., Murphy, B.D. and Bordignon, V. 2008. Adiponectin enhances in vitro development of swine embryos. *Domestic Animal Endocrinology* 35(2):198-207.

In: Proteins Researcher Biographical Sketches …
Editors: H. Z. Wang and M. Tian
ISBN: 978-1-62100-777-7

Chapter 37

BIOGRAPHICAL SKETCH

NAME:	TITLE:	DATE OF BIRTH:
Kouichi Nakagawa	DEPARTMENT OF RADIOLOGICAL LIFE SCIENCES GRADUATE SCHOOL OF HEALTH SCIENCES, HIROSAKI UNIVERSITY,	January 3, 1956

EDUCATION:

Institution and Location	Degree	Year Conferred	Scientific Field
Chemistry, West Virginia University	M. S	1985	
Chemistry, Boston University	Ph. D.	1989	

CONTACT POINTS:

Address: 66-1 Hon-cho, Hirosaki 036-8564, JAPAN

RESEARCH AND PROFESSIONAL EXPERIENCE:

1989 – 1990 Research Associate, Northwestern University
1990 – 1992 Research Associate, University of Denver

Professional Appointments:

2010 ~ Present, Full Professor

Publications during Last Three Years:

1) K. Nakagawa, "Elucidated Lipid Structures of Various Human Stratum Corneum Investigated by EPR Spectroscopy, *Skin Research and Technology*, in press (2011).
2) K. Nakagawa and K. Anzai, "Stratum Corneum Lipid Structure Investigated by EPR Spin-Probe Method: Application of Terpenes," *Lipids*, 45, 1081-1087 (2010). DOI: 10.1007/s11745-010-3479-z
3) K. Nakagawa and K. Anzai, "EPR Investigation of Radical-Production Cross Sections for Sucrose and L-Alanine Irradiated with X-ray and Heavy Ions," *Appl. Magn. Reson.*, 39(3), 285-293 (2010). DOI: 10.1007/s00723-010-0157-5
4) K. Nakagawa, "Electron Paramagnetic Resonance Investigation of Stratum Corneum Lipid Structure, *Lipids*, 45, 91-96 (2010). DOI: 10.1007/s11745-009-3374-7
5) Y. Karakirova, K. Nakagawa, N. D. Yordanov, "EPR and UV spectroscopic investigations of sucrose irradiated with heavy-ion particles," *Radiat. Meas.*, 45, 10-14 (2010). DOI: 10.1016/j.radmeas.2009.07.003
6) K. Nakagawa, "Structure of stratum corneum lipid studied by electron paramagnetic resonance," Chapter 70, *Textbook of Aging Skin*, Miranda A. Farage, Kenneth W. Miller and Howard I. Maibach Eds, Springer-Verlag, 725-733 (2009). ISBN 978-3-540-89655-5
7) K. Nakagawa, "EPR Investigations of Spin-Probe Dynamics in Aqueous Dispersions of a Nonionic Amphiphilic Compound," *J. Am. Oil Chem. Soc.*, 86, 1-17 (2009). [Review article] DOI: 10.1007/s11746-008-1317-8
8) K. Nakagawa, "Electron-Spin Lattice Relaxation Times of Spin Probes in Aqueous Dispersions of a Unique Amphiphilic Compound Obtained by a Saturation Recovery Method," *Bull. Chem. Soc. Jpn.*, 81, 843-846 (2008). DOI: 10.1246/bcsj.81.843
9) K. Nakagawa, N. Ikota, and K. Anzai, "Sucrose and L-alanine radical-production cross section regarding heavy-ion irradiation," *Spectrochimica Acta Part A, Molecular & Biomolecular Spectroscopy*, 69, 1384-1387 (2008). DOI: 10.1016/j.saa.2007.09.041
10) K. Nakagawa, N. Ikota, and Y. Sato, "Heavy-Ion Induced Sucrose and L-α-Alanine Radicals Investigated by Electron Paramagnetic Resonance," *Appl. Magn. Reson.*, 33, 111-116 (2008). DOI: 10.1007/s00723-008-0052-5

In: Proteins Researcher Biographical Sketches …
Editors: H. Z. Wang and M. Tian
ISBN: 978-1-62100-777-7

Chapter 38

BIOGRAPHICAL SKETCH

NAME:	TITLE:	DATE OF BIRTH:
Kenji Osawa	DEPARTMENT OF ORAL AND MAXILLOFACIAL SURGERY, GRADUATE SCHOOL OF MEDICINE, KYOTO UNIVERSITY	June 5th 1976

EDUCATION:

Institution and Location	Degree	Year Conferred	Scientific Field
Department of Oral and Maxillofacial Surgery, Graduate school of Medicine, Kyoto University, Kyoto, Japan	Ph.D.	2007-2011	
Kyusyu Dental College, Fukuoka, Japan	D.D.S.	1997-2003	

CONTACT POINTS:

Address:
#1402 1-7-28 Manazuru, Kitakyusyu, Fukuoka 803-0844, Japan.

RESEARCH AND PROFESSIONAL EXPERIENCE:

2010-Present Research Fellow of the Japan Society for the Promotion of Science (JSPS Research fellow)
2005-2007 Medical Staff in Ako-city Hospital, Department of Oral and Maxillofacial Surgery, Hyogo, Japan
2003-2004 Senior Resident in Oral and Maxillofacial Surgery, Kyoto University Hospital
2002-2003 Junior Resident in Oral and Maxillofacial Surgery, Kyoto University Hospital
2002 Passed the national board of dental examination of Japan

Publications during Last Three Years:

K. Osawa, Y. Okubo, K. Nakao, N. Koyama, K. Bessho: Osteoinduction by repeat plasmid injection of human bone morphogenetic protein-2. J. Gene Med. 2010; 12: 937-944.
N. Koyama, Y Okubo, K Nakao, K,Osawa, K. Bessho: Experimental study of osteoinduction using a new material as a carrier for bone morphogenetic protein-2. Br. J. Oral maxillofac. Surg. 2010; in press.
K. Osawa, Y. Okubo, K. Nakao, N. Koyama, K. Bessho: Osteoinduction by microbubble-enhanced transcutaneous sonoporation of human bone morphogenetic protein-2. J. Gene Med. 2009; 11: 633-641.

In: Proteins Researcher Biographical Sketches … ISBN: 978-1-62100-777-7
Editors: H. Z. Wang and M. Tian

Chapter 39

BIOGRAPHICAL SKETCH

NAME:	TITLE:	DATE OF BIRTH:
Marie-France Palin	RESEARCH SCIENTIST AT AGRICULTURE AND AGRI-FOOD CANADA, SHERBROOKE, QUEBEC, CANADA	

EDUCATION:

Institution and Location	Degree	Year Conferred	Scientific Field
Sherbrooke University, Quebec, Canada	B.Sc.	1984-1987	Biology majoring in Microbiology
Sherbrooke University, Quebec, Canada	M.Sc.	1988-1990	Biochemistry, direct passage to Ph.D. without writing of the M.Sc. thesis
Sherbrooke University, Quebec, Canada	Ph.D.	1991-1996	Biochemistry

CONTACT POINTS:

Address:
Dairy and Swine Research and Development Centre
Agriculture and Agri-Food Canada
2000, College Street, P.O. Box 90, Lennoxville Stn.
Sherbrooke, Quebec, Canada J1M 1Z3

Professional Appointments:

1994-1996: Research associate (BI-01), Agriculture and Agri-Food Canada, Research Centre, Sherbrooke, Quebec, Canada.

1996-1999: Research scientist (SE-RES-01), Agriculture and Agri-Food Canada, Research Centre, Sherbrooke, Quebec, Canada.

1999-2002: Research scientist (SE-RES-02), Agriculture and Agri-Food Canada, Research Centre, Sherbrooke, Quebec, Canada.

2002-2011: Research scientist (SE-RES-03), Agriculture and Agri-Food Canada, Research Centre, Sherbrooke, Quebec, Canada.

2011-now: Research scientist (SE-RES-04), Agriculture and Agri-Food Canada, Research Centre, Sherbrooke, Quebec, Canada.

1998- now: Adjunct professor at the Animal Science Department, Laval University, Quebec, Quebec, Canada.

2000- now: Adjunct professor at the Biology Department, Sherbrooke University, Sherbrooke, Quebec, Canada.

1994: Lecturer, Biology Department, Sherbrooke University, Sherbrooke, Quebec, Canada.

Honors:

2002: Winner of the Pfizer Canada Inc. Young Scientist Award.
Given by the Canadian Society of Animal Sciences.

Publications during Last Three Years:

Total of 24 scientific publications for the last 3 years.

2011

1. Lefèvre, P.L.C., Palin, M.F., Chen, G., Turecki, G. and Murphy, B.D. 2011. Polyamines are implicated in the emergence of the embryo from obligate diapause. *Endocrinology* 152(4):1627-1639.
2. Lefèvre, P.L.C., Palin, M.F., Beaudry, D., Dobias-Goff, M., Desmarais, J.A., Llerena, E.M. and Murphy, B.D. 2011. Uterine Signaling at the Emergence of the Embryo from Obligate Diapause. *American Journal of Physiology - Endocrinology and Metabolism.* 300: E800-808.
3. Lefèvre, P.L.C., Palin, M.F. and Murphy, B.D. 2011. Polyamines on the reproductive landscape. *Endocrine Reviews.* (Accepted with minor revision)
4. Farmer, C., Palin, M.F. and Martel-Kennes, Y. 2011. Impact of a phase-feeding regimen during prepuberty on growth performance, metabolite status and mammary gland development in gilts. *Journal of Animal Science.* (Submitted).

2010

5. Farmer, C., Palin, M.F., Gilani, G.S., Weiler, H., Vignola, M., Choudhary, R.K. and Capuco, A.V. 2010. Dietary genistein stimulates mammary hyperplasia in gilts. *Animal* 4 (3):454-465.
6. Brochu-Gaudreau, K., Rehfeldt, C., Blouin, R., Bordignon, V., Murphy, B.D. and Palin, M.F. 2010. Adiponectin action from head to toe. *Endocrine* 37(1):11-32. *(Invited Review).*
7. Preynat, A., Lapierre, H., Thivierge, M.C., Palin, M.F., Cardinault, N., Matte, J.J., Desrochers, A. and Girard, C.L. 2010. Effects of supplementary folic acid and vitamin B12 on hepatic metabolism of dairy cows according to methionine supply. *Journal of Dairy Science* 93(5):2130-2142.
8. Talbot, G., Roy, C.S., Topp, E., Kalmokoff, M.L., Brooks, S.P.J., Beaulieu, C., Palin, M.F. and Massé, D.I. 2010. Spatial distribution of some microbial trophic groups in a plug-flow-type anaerobic bioreactor treating swine manure. *Water Science and Technology* 61(5):1147-1155.
9. Houde, A.A., Méthot, S., Murphy, B.D., Bordignon, V. and Palin, M.F. 2010. Relationships between backfat thickness and reproductive efficiency of sows: a two year trial on two commercial herds fixing their backfat thickness at breeding. *Canadian Journal of Animal Science* 90(3):429-436.
10. Farmer, C., Palin, M.F. and Hovey, R.C. 2010. Greater milk yield is related to increased DNA and RNA content but not to the mRNA abundance of selected genes in sow mammary tissue. *Canadian Journal of Animal Science* 90(3):379-388.

2009

11. Roy, C.S., Talbot, G., Topp, E., Beaulieu, C., Palin, M.F. and Massé, D.I. 2009. Bacterial community dynamics in an anaerobic plug-flow type bioreactor treating swine manure. *Water Research* 43 (1):21-32.
12. Labrecque, B., Mathieu, O., Bordignon, V., Murphy, B.D. and Palin, M.F. 2009. Identification of differentially expressed genes in a porcine *in vivo* model of adipogenesis using suppression subtractive hybridization. *Comparative Biochemistry and Physiology Part D: Genomics and Proteomics* 4(1):32-44.
13. Preynat, A., Lapierre, H., Thivierge, C.M., Palin, M.F., Matte, J.J., Desrochers, A. and Girard, C.L. 2009. Effects of supplements of folic acid, vitamin B12 and rumen-protected methionine on whole body metabolism of methionine and glucose in lactating dairy cows. *Journal of Dairy Science* 92(2):677-689.
14. Preynat, A., Lapierre, H., Thivierge, C.M., Palin, M.F., Matte, J.J., Desrochers, A. and Girard, C.L. 2009. Influence of methionine supply on the response of lactational performance of dairy cows to supplementary folic acid and vitamin B12. *Journal of Dairy Science* 92(4):1685-1695.
15. Labrecque, B., Beaudry, D., Mayhue, M., Hallé, C., Bordignon, V., Murphy, B.D. and Palin, M.F. 2009. Molecular characterization and expression analysis of the porcine paraoxonase 3 (PON3) gene. *GENE* 443(1-2):110-120.
16. Talbot, G., Roy, C.S., Topp, E., Beaulieu, C., Palin, M.F. and Massé, D.I. 2009. Multivariate statistical analyses of rDNA and rRNA fingerprint data to differentiate microbial communities in swine manure. *FEMS Microbiology Ecology* 70(3):540-552.

2008

17. Campos, D.B., Palin, M.F., Bordignon, V. and Murphy, B.D. 2008. The "beneficial" adipokines in reproduction and fertility. *International Journal of Obesity (London)* 32(2):223-231. *(Review)*
18. Talbot, G., Topp, E., Palin, M.F. and Massé, D.I. 2008. Evaluation of molecular methods used for establishing the interactions and functions of microorganisms in anaerobic bioreactors. *Water Research* 42(3):513-537. *(Review)*
19. Vallée, M., Aiba, K., Piao, Y., Palin, M.F., Ko, M.S. and Sirard, M.A. 2008. Comparative analysis of oocyte transcript profiles reveals a high degree of conservation among species. *Reproduction* 135(4):439-448.
20. Houde, A.A., Murphy, B.D., Mathieu, O., Bordignon, V. and Palin, M.F. 2008. Characterization of swine adiponectin and adiponectin receptor polymorphisms and their association with reproductive traits. *Animal Genetics* 39(3):249-57.
21. Palin, M.F., Labrecque, B., Beaudry, D., Mayhue, M., Bordignon, V. and Murphy, B.D. 2008. Visfatin expression is not associated with adipose tissue abundance in the porcine model. *Domestic Animal Endocrinology* 35(1):58-73.
22. Chappaz, E., Albornoz, M.S., Campos, D., Che, L., Palin, M.F., Murphy, B.D. and Bordignon, V. 2008. Adiponectin enhances in vitro development of swine embryos. *Domestic Animal Endocrinology* 35(2):198-207.
23. Petit, H.V. and Palin, M.F. 2008. *Letter to the Editor*: A Response to the Comments of Rastani and Kertz. *Journal of Dairy Science* 91:2534.
24. Farmer, C. and Palin, M.F. 2008. Feeding flaxseed to sows during late-gestation and lactation affects mammary development but not mammary expression of selected genes in their offspring. *Canadian Journal of Animal Science* 88(4):585-590.

In: Proteins Researcher Biographical Sketches … ISBN: 978-1-62100-777-7
Editors: H. Z. Wang and M. Tian

Chapter 40

BIOGRAPHICAL SKETCH

NAME:	TITLE:	DATE OF BIRTH:
Joanna E. Pankiewicz		

EDUCATION:

Institution and Location	Degree	Year Conferred	Scientific Field
Collegium Medicum Jagiellonian Univ. Cracow, Poland	MD	1994	Medicine
Collegium Medicum Jagiellonian Univ. Cracow, Poland	PhD	2001	Clinical Biochemistry
New York University, School of Medicine, New York	Postdoctoral fellowship	2003-2007	Neuroscience

A. Positions and Honors

Positions and Employment

1994-1995: Internship in Medicine, Surgery, Obstetrics, and Pediatrics, University Hospital, Collegium
Medicum, Jagiellonian University, Cracow, Poland

1995-200: PhD program, Dept. of Clinical Biochem., Collegium Med. Jagiellonian Univ. Cracow, Poland

1996-2002: Neurology Resident, Dept. of Neurology, Collegium Med. Jagiellonian Univ. Cracow, Poland

2003-2007: Associate Research Scientist, Dept. of Neurology, New York University School of Medicine
2007: Assistant Professor (Research), Dept. of Neurology, New York University School of Medicine
2008: Assistant Professor (Research), Dept. of Pharmacology, New York Univ. School of Medicine

B. Selected Peer-reviewed publications (in chronological order)

1) L.Partyka, A. Dembinska-Kiec, J. Pankiewicz, D.Dudek, J. Hartwich, A. Siedlacki, E. Pawlus, M. Marquez-Perez, J. Torres-Herrera. An inhibitory effect of camonagrel-a new tromboxane synthase inhibitor, on Pselectin mediated platelet/PMN adhesion. Platelets 1996, 7, 169-172
A. Szczudlik, A. Deminska-Kiec, A. Słowik, G. Zwolinska, J. Pankiewicz, B. Tomik, U. Wyrwicz-Petkow, J. Furgal, A. Zdzienicka. Stress hormones response and and transient hyperglycemia in acute stroke in nondibetic patients. Acta Angiologica, 1996, 2, 4:255-261
2) Deminska-Kiec, K. Kawecka-Jaszcz, M. Kwasniak, I. Guevara, J. Pankiewicz, J. Iwanejko, A. Zdzienicka, A. Stochmal, I. Leszczynska-Gołabek. Apo E isoforms, insulin output and plasma lipid levels in essential hypertension. European Journal of Clinical Investigation 1998:28;95-99
3) Guevara, J. Iwanejko, A. Dembinska-Kiec, J. Pankiewicz, A. Wanat, A. Polus, I. Gołabek, S. Bartus, M. Malczewska-Malec, A. Szczudlik. Determination of nitrate/nitrite in human biological material by simple Griess reaction. Clinica Chimica Acta 1998:274;177-188.
4) I Leszczyńska-Gołąbek, Ł. Partyka J. Pankiewicz, B. Piwońska, A. Demińska-Kieć, , A. Sieczkowski, A. Cencora, M. Marquez, J. Torres-Herrera. Influence of a novel thromboxane synthetase inhibitor-camongrel an the clinical status of patients with peripheral arterial disease. Acta Angiologica 1999;5 (1/2);27-34.
5) Leszczynska-Golabek, L. Partyka, J. Pankiewicz, B. Piwonska, A. Dembinska-Kiec, A. Sieczkowski, A. Cencoera, M. Marquez, J. Toress-herera. Influence of a novel thromboxane synthase inhibitor-camonagrel on clinical staaatus of patietns with peripheral arterial disease. Acta Angiologica 1999:5;27-34.
6) Pankiewicz, A. Dembinska-Kiec, A. Slowik, M. Rudzinska, A. Szczudlik. The role of statins in prevention of ischaemic stroke. Przegląd Lekarski 2000;57 (3);406-410.
7) W. Turaj, A. Slowik, J. Pankiewicz, T. Iskra, M. Rudzinska, A. Szczudlik. The prognostic significance of microalbuminuria in non-diabetic acute stroke patients. Med. Sci Monit 2001:7 (5);989-994.
8) T. Iskra, W Turaj, J.Hartwich, A. Slowik, J. Pankiewicz, A. Dembinka-Kiec. LDL phenotype A and B in ischemic stroke. Przegląd Lekarski 2002;59 (1);7-10.
9) Dziedzic T, Wybranska I, Dembinska-Kiec A, Klimokowicz A, Slowik A, Pankiewicz J, Zdzienicka A, Szczudlik A. Dexamethasone inhibits TNF-alfa synthesis more effectively in Alzheimer's disease patients than in healthy individuals. Dement Geriatr Cogn Disord. 2003;16(4):283-6.
10) Slowik A, Turaj W, Pankiewicz J, Dziedzic T, Szermer P, Szczudlik A. Hypercortisolemia in acute stroke is related to the inflammatory response. J of Neurolo Sci. 2002;196, 27-32.

11) Sadowski M, Pankiewicz J, Scholtzova H, Li Y-S, Quartermain D, Wisniewski T Links between the pathology of Alzheimer's disease and vascular dementia. Neurochemical Res. (2004) 29(6):1257-66.
12) Sadowski M, Pankiewicz J, Scholtzova H, Quartermain D, Jensen CH, Duff K, Nixon RA, Gruen RJ, Wisniewski T. Amyloid-β deposition is associated with decreased hippocampal glucose metabolism and spatial memory impairment in APP/PS1 mice. J Neuropath Exp Neurol (2004) 63(5) 418-428.
13) Sadowski M, Pankiewicz J, Scholtzova H, Ripellino J, Schmidt SD, Mathews PM, Sigurdsson EM, Holtzman D, Wisniewski T. A synthetic peptide blocking the apolipoprotein E/β-amyloid binding mitigates β-amyloid toxicity and fibril formation in vitro and reduces β-amyloid plaques in transgenic mice. Am. J. Pathol. (2004) 165(3): 937-948.
14) Sadowski M, Pankiewicz J, Scholtzova H, Tsai J, Carp RI, Meeker CH, Gan WB, Klunk WE, Wisniewski T.Targeting priom amyloid deposits in vivo. J Neuropath Exp Neurol (2004) 63(7):775-784.
15) Goni,F., Knudsen,E., Schreiber,F., Scholtzova,H., Pankiewicz,J., Carp,R., Meeker,H.C., Rubenstein,R., Brown,D.R., Sy,M.S., Chabalgoity,J.A., Sigurdsson,E.M., and Wisniewski,T., Mucosal vaccination delays or prevents prion infection via an oral route, Neuroscience, 133 (2005) 413-421.
16) Pankiewicz J, Prelli F, Sy M-S, Kascsak R, Kascsak R, Spinner D, Carp R, Meeker H, Sadowski M, Wisniewski T, Clearance and Prevention of Prion Infection in Cell Culture by Anti-PrP Antibodies. Eur. J. Neurosc. 23 (2006) 2635-2647
17) Sadowski M, Pankiewicz J, Mehta P, Sholtzova H, Wen P, Prelli F, Quartermain D, Wisniewski T. Blocking the apolipoproteinE/amyloid-â Interaction as a potential therapeutic approach for Alzheimer's disease. PNAS 103 (2006) 18787-18792.
18) Sadowski M[*] Pankiewicz J, Prelli F, Scholtzova H, Spinner DS, Kascsak R, Kascsak R, , Wisniewski T[*], Anti-PrP Mab 6D11 Suppresses PrP^{Sc} Replication in Prion Infected Myeloid Precursor Line FDC-P1/22L and in the Lymphoreticular System In Vivo. *Neurobiol Dis* 2009 Neurobiol Dis. 2009 May;34(2):267-78

C. Research Support:

NIRG Pankiewicz (PI) 08/01/09-07/31/11
Alzheimer's Association
<u>Passive Immunization for Prion Infections</u>

The major goal of this grant is to investigate passive immunization with anti-prion monoclonal antibodies in mice models of prion infection.

Role: Principal Investigator

In: Proteins Researcher Biographical Sketches … ISBN: 978-1-62100-777-7
Editors: H. Z. Wang and M. Tian

Chapter 41

BIOGRAPHICAL SKETCH

NAME:	**TITLE:**	**DATE OF BIRTH:**
Ãrika Cristina Pavarino	**PHD. FACULDADE DE MEDICINA DE SÃO JOSE DO RIO PRETO (FAMERP), SÃO PAULO, BRAZIL**	31/07/1969

CONTACT POINTS:

Address: Avenida Brigadeiro Faria Lima, 5416, São José © do Rio Preto, São Paulo, Brazil

RESEARCH AND PROFESSIONAL EXPERIENCE:

Molecular Biology and Genetics

Professional Appointments:

Professor

Publications during Last Three Years:

1. Galbiatti, a. L. S.; ruiz, m. T.; rezende-pinto, d.; raposo, l. S.; manã□glia, j. V.; pavarino, e. C.; goloni-bertollo, e. M. A80g polymorphism of reduced folate carrier 1 (rfc1) gene and

head and neck squamous cell carcinoma etiology in Brazilian population. Molecular Biology Reports, v. 38, p. 1071-1078, 2011.

2. Silva, L. M. R. B.; Silva, J. N. G.; Galbiatti, A.L.S.; Succi, M. ; Ruiz, M. T.; Raposo, L. S.; Maniglia, J. V.; Pavarino-Bertelli, E.C.; Goloni-Bertollo, E. M. CarcinogÃªnese de cabeÃ§a e pescoÃ§o: impacto do polimorfismo Mthfd1 G1958A. AMB (SÃ£o Paulo) (Cessou em 1991. Cont. ISSN 0104-4230 Revista da AssociaÃ§Ã£o MÃ©dica Brasileira. (1992. Impresso)), v. 57, p. 194-199, 2011.
3. Marucci, g. H.; zampieri, b. L.; biselli, j. M.; valentim, s.; goloni-bertollo, e. M.; eberlin, m. N.; haddad, r.; riccio, m. F.; vannucchi, h.; carvalho, v. M. ; pavarino, e.c. Polymorphism c1420t of serine hydroxymethyltransferase gene on maternal risk for down syndrome. Molecular biology reports, v. 38, p. 2869, 2011.
4. Cury, n. M.; russo, a.; galbiatti, a.l.s.; ruiz, m. T.; raposo, l. S.; maniglia, j. V.; goloni-bertollo, e. M.; pavarino, e.c. Polymorphisms of the cyp1a1 and cyp2e1 genes in head and neck squamous cell carcinoma risk. Molecular biology reports, v. 38, p. 2869, 2011.
5. Ruiz, m. T.; biselli, p. M.; maniglia, j. V.; pavarino-bertelli, e.c.; goloni-bertollo, e. M. Vascular endothelial growth factor genetic variability and head and neck cancer in a brazilian population. Brazilian journal of medical and biological research (impresso), v. 43, p. 127-133, 2010.
6. Mendes, c. C.; biselli, j. M.; zampieri, b. L.; goloni-bertollo, e. M.; eberlin, m. N.; haddad, r.; riccio, m. F.; vannucchi, h.; carvalho, v. M.; pavarino-bertelli, e.c. 19-base pair deletion polymorphism of dihydrofolate reductase (dhfr) gene: maternal risk for down syndrome and folate metabolism. Sã£o paulo medical journal (impresso), v. 128, p. 215-218, 2010.
7. Leme, c. V. D.; raposo, l. S.; ruiz, m. T.; biselli, j. M.; galbiatti, a. L. S.; maniglia, j. V.; pavarino-bertelli, e. C.; goloni-bertollo, e. M. Anã¡lise dos genes gstm1 e gstt1 em pacientes com cã¢ncer de cabeã§a e pescoã§o. Revista da associaã§ã£o mã©dica brasileira (1992. Impresso), v. 56, p. 299-303, 2010.
8. Rodrigues, J. O.; Galbiatti, A. L. S.; Ruiz, M. T.; RAPOSO, L. S.; Maniglia, J. V.; Bertelli, E. C. P.; Goloni-Bertollo, E. M. Polimorfismo do Gene Metilenotetrahidrofolato Redutase (MTHFR) e o risco de Carcinoma Espinocelular de cabeã§a e pescoã§o. Revista Brasileira de Otorrinolaringologia (Online), v. 76, p. 776-782, 2010.
9. Ruiz, m. T.; balachi, j. F.; fernandes, r. A.; manã• glia, j. V.; bertelli, e. C. P.; goloni-bertollo, e. M. Anã¡lise do gene tax1bp1 em pacientes com cã¢ncer de cabeã§a e pescoã§o. Revista brasileira de otorrinolaringologia (impresso), v. 76, p. 193-198, 2010.
10. Galbiatti, A.L.S.; Ruiz, M. T.; Biselli-Chicote, P. M.; Raposo, L. S.; Maniglia, J. V.; Pavarino-Bertelli, E.C.; Pavarino-Bertelli, E.C.; Goloni-Bertollo, E. M. 5-methyltetrahydrofolate-homocysteine methyltransferase gene polymorphism (MTR) and risk of head and neck cancer. Brazilian Journal of Medical and Biological Research (Impresso), v. 43, p. 124-225, 2010.
11. Galbiatti, a.l.s.; ruiz, m. T.; raposo, l. S.; maniglia, j. V.; pavarino-bertelli, e. C.; goloni-bertollo, e. M. The association between cbs 844ins68 polymorphism and head and neck squamous cell carcinoma risk a case-control analysis. Archives of medical science, v. 6, p. 772-779, 2010.
12. Zuccari, d. A. P.; castro, r.; gomes, a. F. G.; mancini, u. M.; frade, c. S.; pivaro, l. R.; carmona, j.; terzian, a. C. B.; ruiz, c. M.; goloni-bertollo, e. M.; pavarino-bertelli, e. C.;

tajara, e. H. The maspin expression in canine mammary tumors: an immunohistochemical and molecular study. Pesquisa veterinã¡ria brasileira, v. 29, p. 163-173, 2009.

13. Biselli, j. M.; zampieri, b. L.; silva, a. F. A.; goloni-bertollo, e. M.; haddad, r.; carvalho, v. M.; pavarino-bertelli, e. C. Double aneuploidy (48,xxy,+21) of maternal origin in a child born to a 13-year-old mother: evaluation of the maternal folate metabolism. Journal of genetic counseling, v. 20, p. 225-234, 2009.
14. Biselli, j. M.; goloni-bertollo, e. M.; ruiz, m. T.; pavarino-bertelli, e.c. Cytogenetic profile of down syndrome cases seen by a general genetics outpatient service in brazil. Down's syndrome. Rresearch and practice, v. Online, p. 10.3104, 2009.
15. Biselli, p. M.; guerzoni, a. R.; godoy, m. F.; eberlin, m. N.; haddad, r.; carvalho, v. M.; vannucchi, h.; pavarino-bertelli, e.c.; goloni-bertollo, e. M. Genetic polymorphisms involved in folate metabolism and concentrations of methylmalonic acid. Journal of thrombosis and thrombolysis, v. 1, p. 1-1, 2009.
16. Biselli, p. M.; guerzoni, a. R.; goloni-bertollo, e. M.; godoy, m. F.; abou-chahla, j. A. B.; pavarino-bertelli, e.c. Mthfr genetic variability on coronary artery disease development. Revista da associaã§ã£o mã©dica brasileira (1992. Impresso), v. 55, p. 274-278, 2009.
17. Pavarino-bertelli, e.c.; biselli, j. M.; bonfim, d.; goloni-bertollo, e. M. Clinical profile of children with down syndrome treated in a genetics outpatient service in the southeast of brazil.. Revista da associaã§ã£o mã©dica brasileira (1992. Impresso), v. 55, p. 547-552, 2009.
18. Guerzoni, a. R.; biselli, p. M.; godoy, m. F.; souza, d. R. S.; haddad, r.; eberlin, m. N.; pavarino-bertelli, e.c.; goloni-bertollo, e. M. Homocysteine and polymorphisms of mthfr and vegf genes: impact in coronary artery disease. Arquivos brasileiros de cardiologia (impresso), v. 92, p. 249-254, 2009.

In: Proteins Researcher Biographical Sketches … ISBN: 978-1-62100-777-7
Editors: H. Z. Wang and M. Tian

Chapter 42

BIOGRAPHICAL SKETCH

NAME:	TITLE:	DATE OF BIRTH:
Vassilios Raikos	**DEPARTMENT OF CHEMISTRY, LABORATORY OF PHYSICAL CHEMISTRY, UNIVERSITY OF PATRAS, GREECE**	22/07/1976

EDUCATION:

Institution and Location	Degree	Year Conferred	Scientific Field
University of Leeds, UK	BSc		Human Genetics
University of Nottingham, UK	MSc		Applied Biomolecular Technology
Heriot-Watt University, UK)	PhD		Food Science

CONTACT POINTS:

Address: Department of Chemistry, Laboratory of Physical Chemistry, University of Patras, PC 26504, Patras, Greece

RESEARCH AND PROFESSIONAL EXPERIENCE:

- Post-Doctoral Research associate (Department of Chemistry, Laboratory of Physical Chemistry University of Patras, Patras, Greece, 2007-today)
- Teaching associate (Department of Food Technology, Technological Institute of Kalamata, Kalamata, Greece, 2008-today)

Publications during Last Three Years:

1) The use of sedimentation field-flow fractionation in the size characterization of bovine milk fat globules as affected by heat treatment (2009) Vassilios Raikos, John Kapolos, Lambros Farmakis, Athanasia Koliadima and George Karaiskakis, *Food Research International*, **42 (5-6)**, pp. 659-665
2) Effect of heat treatment on milk protein functionality at emulsion interfaces. A review (2010) Vassilios Raikos, *Food Hydrocolloids*, **24 (4)**, pp. 259-265

In: Proteins Researcher Biographical Sketches … ISBN: 978-1-62100-777-7
Editors: H. Z. Wang and M. Tian

Chapter 43

BIOGRAPHICAL SKETCH

NAME:	TITLE:	DATE OF BIRTH:
Gu Seob Roh	DEPARTMENT OF ANATOMY, SCHOOL OF MEDICINE, GYEONGSANG NATIONAL UNIVERSITY	72.03.13.

EDUCATION:

Institution and Location	Degree	Year Conferred	Scientific Field
College of Medicine, Gyeongsang National University	M.D.	1998	
Department of Anatomy, College of Medicine, Gyeongsang National University	M.S.	2000	
Department of Anatomy, College of Medicine, Gyeongsang National University	Ph.D.	2002	

CONTACT POINTS:

Address:
Department of Anatomy and Neurobiology, School of Medicine, Gyeongsang National University, 816 Beongil 15 Jinju-daero, Jinju, Gyeongnam 660-290, Republic of Korea

RESEARCH AND PROFESSIONAL EXPERIENCE:

Seizure, Neurobiology, Neuroinflammation, Neurodegeneration, Diabetes, Asthma

Professional Appointments:

2011 – Present Associate Professor
Department of Anatomy,
School of Medicine, Gyeongsang National University
2007 – 2011 Assistant Profrssor
Department of Anatomy,
School of Medicine, Gyeongsang National University
2005 – 2007 Full-time lecturer
Department of Anatomy,
School of Medicine, Gyeongsang National University
2003 – 2005 Post-Doctoral Research Associate
Center for Genome Science, National Institute of Health, Korea Centers for Disease Control and Prevention, Republic of Korea
2002 – 2003 Post-Doctoral Research Associate
Department of Forensic Medicine, National Institute of Scientific Investigation, Republic of Korea
1998 – 2002 Graduate Research Assistant
Department of Anatomy,
College of Medicine, Gyeongsang National University

Honors:

2010 Marquis Who's Who

Publications during Last Three Years:

1. Ku BM, Lee YK, Jeong JY, Ryu J, Choi J, Kim JS, Cho YW, Roh GS, Kim HJ, Cho GJ, Choi WS, Kang SS. Caffeine inhibits cell proliferation and regulates PKA/GSK3 pathways in U87MG human glioma cells. Mol Cells. 2011 Mar;31(3):275-9.
2. Lee DH, Jeong JY, Kim YS, Kim JS, Cho YW, Roh GS, Kim HJ, Kang SS, Cho GJ, Choi WS. Ethanol down regulates the expression of myelin proteolipid protein in the rat hippocampus. Anat Cell Biol. 2010 Sep;43(3):194-200.
3. Jeong EA, Jeon BT, Kim JB, Kim JS, Cho YW, Lee DH, Kim HJ, Kang SS, Cho GJ, Choi WS, Roh GS. Phosphorylation of 14-3-3 at serine 58 and neurodegeneration following kainic acid-induced excitotoxicity. Anat Cell Biol. 2010 Jun;43(2):150-6.

4. Kim YS, Choi MY, Kim YH, Jeon BT, Lee DH, Roh GS, Kang SS, Kim HJ, Cho GJ, Choi WS. Protein kinase Cdelta is associated with 14-3-3 phosphorylation in seizure-induced neuronal death. Epilepsy Res. 2010 Nov;92(1):30-40.
5. Kim G, Lee Y, Jeong EY, Jung S, Son H, Lee DH, Roh GS, Kang SS, Cho GJ, Choi WS, Kim HJ. Acute stress responsive RGS proteins in the mouse brain. Mol Cells. 2010 Aug;30(2):161-5.
6. Kim CW, Tak HM, Kim GT, Mun YJ, Jeon BT, Kim HJ, Roh GS, Han J, Kang D. H2O2-induced up-regulation of CatSper3 in mouse brain. Mol Reprod Dev. 2010 Aug;77(8):650.
7. Ku BM, Lee YK, Ryu J, Jeong JY, Choi J, Eun KM, Shin HY, Kim DG, Hwang EM, Yoo JC, Park JY, Roh GS, Kim HJ, Cho GJ, Choi WS, Paek SH, Kang SS. CHI3L1 (YKL-40) is expressed in human gliomas and regulates the invasion, growth and survival of glioma cells. Int J Cancer. 2011 Mar 15;128(6):1316-26.
8. Roh GS, Yi CO, Cho YJ, Jeon BT, Nizamudtinova IT, Kim HJ, Kim JH, Oh YM, Huh JW, Lee JH, Hwang YS, Lee SD, Lee JD. Anti-inflammatory effects of celecoxib in rat lungs with smoke-induced emphysema. Am J Physiol Lung Cell Mol Physiol. 2010 Aug;299(2):L184-91.
9. Lee DH, Jeon BT, Jeong EA, Kim JS, Cho YW, Kim HJ, Kang SS, Cho GJ, Choi WS, Roh GS. Altered expression of sphingosine kinase 1 and sphingosine-1-phosphate receptor 1 in mouse hippocampus after kainic acid treatment. Biochem Biophys Res Commun. 2010 Mar 12;393(3):476-80.
10. Kim JH, Lee SS, Jung MH, Yeo HD, Kim HJ, Yang JI, Roh GS, Chang SH, Park DJ. N-acetylcysteine attenuates glycerol-induced acute kidney injury by regulating MAPKs and Bcl-2 family proteins. Nephrol Dial Transplant. 2010 May;25(5):1435-43.
11. Cho YJ, Han JY, Lee SG, Jeon BT, Choi WS, Hwang YS, Roh GS, Lee JD. Temporal changes of angiopoietins and Tie2 expression in rat lungs after monocrotaline-induced pulmonary hypertension. Comp Med. 2009 Aug;59(4):350-6.
12. Kim JS, Son D, Choi TH, Han K, Kim JH, Cho HM, Kim WH, Kim SH, Kim NG, Lee KS, Hwang OH, Roh GS, Park J. Interferon alpha-2a reduces early erythema after full-thickness skin graft in the pig. Dermatol Surg. 2009 Oct;35(10):1514-24.
13. Lee KW, Lee DH, Son H, Kim YS, Park JY, Roh GS, Kim HJ, Kang SS, Cho GJ, Choi WS. Clusterin regulates transthyretin amyloidosis. Biochem Biophys Res Commun. 2009 Oct 16;388(2):256-60.
14. Joo Y, Choi KM, Lee YH, Kim G, Lee DH, Roh GS, Kang SS, Cho GJ, Choi WS, Kim HJ. Chronic immobilization stress induces anxiety- and depression-like behaviors and decreases transthyretin in the mouse cortex. Neurosci Lett. 2009 Sep 18;461(2):121-5.
15. Jeon BT, Shin HJ, Kim JB, Kim YK, Lee DH, Kim KH, Kim HJ, Kang SS, Cho GJ, Choi WS, Roh GS. Adiponectin protects hippocampal neurons against kainic acid-induced excitotoxicity. Brain Res Rev. 2009 Oct;61(2):81-8.
16. Jeon BT, Lee DH, Kim KH, Kim HJ, Kang SS, Cho GJ, Choi WS, Roh GS. Ketogenic diet attenuates kainic acid-induced hippocampal cell death by decreasing AMPK/ACC pathway activity and HSP70. Neurosci Lett. 2009 Mar 27;453(1):49-53.
17. Lee JY, Jeon BT, Shin HJ, Lee DH, Han JY, Kim HJ, Kang SS, Cho GJ, Choi WS, Roh GS. Temporal expression of AMP-activated protein kinase activation during the kainic acid-induced hippocampal cell death. J Neural Transm. 2009 Jan;116(1):33-40.

ISBN: 978-1-62100-777-7

Chapter 44

BIOGRAPHICAL SKETCH

NAME:	TITLE:	DATE OF BIRTH:
Krisna Rungruangsak-Torrissen	**DR.** Institute of Marine Research, Ecosystem Processes Research Group, Matre Research Station, N-5984 Matredal, Norway	1 January 1950

EDUCATION:

Institution and Location	Degree	Year Conferred	Scientific Field
University of Bergen, Bergen, Norway	Dr. Philos.		Fisheries and Marine Biology
Faculty of Graduate Studies, Mahidol University, Bangkok, Thailand	M.Sc.		Biochemistry
Faculty of Science, Chulalongkorn University, Bangkok, Thailand	B.Sc.		Food Technology

CONTACT POINTS:

Address: Haugsværlia 14, N-5983 Haugsvær, Norway

RESEARCH AND PROFESSIONAL EXPERIENCE:

Food utilization and performance qualities of growth and maturation of aquatic animals in both aquaculture and natural ecosystem using unique combinations of different biochemical techniques (digestive efficiency and protein digestion, trypsin isozymes, protein synthesis and turnover, white muscle quality, oocyte quality, dietary quality assessment)

Professional Appointments:

Principal Research Scientist

Publications during Last Three Years:

Rungruangsak-Torrissen K., U. Kovitvadhi, J. Siruntawineti, E. Trenet, A. Engkagul, Y. Aeidnoie, K. Choowongkomon, S. Unajak, P. Meeswad, J. Sunthornchot, S. Kovitvadhi and P. Preprame. 2010. Development of suitable feed for commercial production of Nile tilapia, *Oreochromis niloticus*. Kasetsart University Technical Report 2010, 184 pp. (in Thai). Biochemical Research Unit for Feed Utilization Assessment, Kasetsart University, Bangkok, Thailand.

Thongprajukaew K., U. Kovitvadhi, A. Engkagul and K. Rungruangsak-Torrissen. 2010. Characterization and expression levels of protease enzymes at different developmental stages of Siamese fighting fish (*Betta splendens* Regan, 1910). Kasetsart J. (Nat. Sci.) 44, 411–423.

Thongprajukaew K., U. Kovitvadhi, A. Engkagul and K. Rungruangsak-Torrissen. 2010. Temperature and pH characteristics of amylase and lipase at different developmental stages of Siamese fighting fish (*Betta splendens* Regan, 1910). Kasetsart J. (Nat. Sci.) 44, 210–219.

Khrueanet, W., U. Kovitvadhi, A. Engkagul, S. Kovitvadhi and K. Rungruangsak-Torrissen. 2009. Characterization of digestive enzymes of adult freshwater pearl mussel *Chamberlainia hainesiana* (Lea, 1856). KKU Sci. J. 37 (Supplement), 11–21.

Rungruangsak-Torrissen K., L.H. Stien, B.S. Daae, T. Vågseth, G.B. Thorsheim, D. Tobin and O. Ritora. 2009. Different dietary levels of protein to lipid ratio affected digestive efficiency, skeletal growth, and muscle protein in rainbow trout families. Scholarly Research Exchange, vol. 2009, Article ID 709529, doi:10.3814/2009/709529.

Rungruangsak-Torrissen K., J. Sunde, A.E. Berg, U. Nordgarden, P.G. Fjelldal and F. Oppedal. 2009. Digestive efficiency, free amino acid pools and quality of growth performance in Atlantic salmon (*Salmo salar* L.) affected by light regimes and vaccine types. Fish Physiol. Biochem. 35, 255–272.

Supannapong P., T. Pimsalee, T. A-komol, A. Engkagul, U. Kovitvadhi, S. Kovitvadhi and K. Rungruangsak-Torrissen. 2008. Digestive enzymes and *in vitro* digestibility of different species of phytoplankton for culture of the freshwater pearl mussel, *Hyriopsis* (*Hyriopsis*) *bialatus*. Aquacult. Int. 16, 437–453.

Chamchuen, P., A. Engkakul, U. Kovitvadhi, K. Rungruangsak-Torrissen and B. Pratoomchat. 2008. Temperature and pH characteristics of digestive enzymes in blue swimming crab larvae, (*Portunus pelagicus*). 10 pp. Proceedings of the 9th National Grad Research Conference, 14–15 March 2008, Burapha University, Bangsaen, Chonburi, Thailand.

Kause, A., L.H. Stien, K. Rungruangsak-Torrissen, O. Ritola, K. Ruohonen and A. Kiessling. 2008. Image analysis as a tool to facilitate selective breeding of quality traits in rainbow trout. Livestock Sci. 114, 315–324.

ISBN: 978-1-62100-777-7

Chapter 45

BIOGRAPHICAL SKETCH

NAME:	TITLE:	DATE OF BIRTH:
Norma Silvia Sánchez	DEPARTAMENTO DE GENETICA MOLECULAR, INSTITUTO DE FISIOLOGIA CELULAR, UNIVERSIDAD NACIONAL AUTONOMA DE MEXICO.	AUGUST 25TH, 1964

EDUCATION:

Institution and Location	Degree	Year Conferred	Scientific Field
School of Chemistry, National University of Mexico, UNAM. Grades: 9.44/ 10.00	Bachelor Degree	1983-1987	Chemistry Pharmaceutics and Biology (Speciality in Microbiologic Biochemistry).

CONTACT POINTS:

Address: Departamento de Genética Molecular, Instituto de Fisiología Celular, circuito Exterior s/n Ciudad Universitaria, CP 04510, México, D.F., México.

RESEARCH AND PROFESSIONAL EXPERIENCE:

16 years as Academic Technician at the Cellular Physiology Institute at the UNAM; 11 published articles in International indexed Journals, 4 as first author, 3 articles in National Journals, 16 acknowledgements in articles published in International indexed journals, 31 participations in National Meetings, 14 participations in International Meetings, 10 courses: Molecular Biology, Radiological Security, Proteomics, Postgraduate Topics, etc. I had a research stay of 3 months at the Centre National de la Recherche Scientifique, Bordeaux, France on 2002. Microbiology and Microbial Physiology Teacher at the Chemistry School of UNAM.

Professional Appointments:

- Quality control laboratory assistant, BECCO Industrial, Tulpetlac, Edo. de México, september- december, 1987.
- Microbiology Teacher. School of Chemistry, UNAM. May, 1990 to March, 1993.
- Associated Technician, Institute of Cellular Physiology, UNAM, Dr. Antonio Peña Díaz laboratory, Microbiology Department. August 1st , 1988 to June 15th, 1993.
- Quality Control Laboratory Manager, ?Procesadora de Ingredientes, SA de CV.?, Calle 2 No. 2725, Col. Ferrocarril, Guadalajara, Jalisco., CP 44440. March - august, 1994 and may, 1997 - march, 1998.
- Associated Technician, Institute of Cellular Physiology, UNAM, Dr. Salvador Uribe Laboratory. Biochemistry Department, from october 16th, 1999- september, 15th, 2003.
- Titular Academic Technician. Institute of Cellular Physiology, UNAM, Dr. Antonio Peña Díaz Laboratory. Molecular Genetics Department. From september, 16th, 2003 and continuing.
- Microbial Physiology Teacher. School of Chemistry, UNAM. From august 2007 and continuing.

Honors:

- Cum Laude, for Bachelor´s Thesis Defense.
- First Place . Bachelor´s Thesis National Contest. ?PREMIO BIOQUIMIA A LA MEJOR TESIS 1989?. Mexican Clinical Biochemistry Association. 1989.
- Permanent Member of the Mexican Biochemistry Society. 2004
- Level C (of D levels. D is the higher level) of the UNAM´s Program of Stimuli to the Academic Productivity. From 2003 and continuing.

Publications during Last Three Years:

- Effects of salts on aerobic metabolism of *Debaryomyces hansenii.* Norma Silvia Sánchez, Roberto Arreguín, Martha Calahorra and Antonio Peña. FEMS Yeast Research (2008) 8:1303-1312.
- Effects of amiodarone on K^{+}, internal pH and Ca^{2+} homeostasis in *Saccharomyces cerevisiae* Antonio Peña, Martha Calahorra, Bertha Michel, Jorge Ramírez and Norma Silvia Sánchez. FEMS Yeast Research. (2009) Vol. 9: 832-848.
- Activation of fermentation by salts in *Debaryomyces hansenii.* Martha Calahorra, Norma Silvia Sánchez and Antonio Peña. FEMS Yeast Research. (2009) Vol. 9: 1293?1301.
- Estimation of the electric plasma membrane potential difference in yeast with fluorescent dyes: comparative study of methods. Antonio Peña, Norma Silvia Sánchez, Martha Calahorra. Journal of Bioenergetics and Biomembranes. (2010) 42:419-432.
- Ketoconazole and miconazole alter potassium homeostasis in Saccharomyces cerevisiae. Martha Calahorra, Carlos Lozano, Norma S Sánchez, Antonio Peña. Biochimica et Biophysica Acta, Biomembranes. (2011) 1808: 433?445
- Changes on the functionality of plasmatic membrane of *Rhizopus stolonifer* by addition of chitosan. Josam Vega-Pérez, Ana Niurka Hernández-Lauzardo , Miguel Gerardo Velázquez-del Valle, Norma Silvia Sánchez- Sánchez, Antonio Peña and Guadalupe Guerra-Sánchez. Journal of Phytopathology, (2011). Accepted for publication.

In: Proteins Researcher Biographical Sketches ...
Editors: H. Z. Wang and M. Tian

ISBN: 978-1-62100-777-7

Chapter 46

BIOGRAPHICAL SKETCH

NAME:	TITLE:	DATE OF BIRTH:
Juan Francisco Santibanez	**PHD.** Laboratory for Experimental Hematology. Institute for Medical Research.	January 29, 1964

EDUCATION:

Institution and Location	Degree	Year Conferred	Scientific Field
Autonoma University of Madrid, Spain	PhD	1999	Biological Science
University of Santiago, Chile	B.Sc.	1990	Biochemistry

CONTACT POINTS:

Address:
Dr. Subotica #4, 11000, Belgrade, Serbia.

RESEARCH AND PROFESSIONAL EXPERIENCE:

Present position: Professor of Research, Laboratory for Experimental Hematology, Institute for Medical Research, University of Belgrade.

Previous positions:

-1989-1990, Bsc thesis (Biochemistry). Facultad de Ciencias Químicas y Farmacéuticas Universidad de Chile. Santiago, Chile, Dr. Mario Sapag-Hagar, Supervisor
-1991-1996, Co-investigator, Cellular Biology Unit. Institute of Nutrition and Food Technology (INTA), Universidad de Chile, Santiago, Chile. Dr. Jorge Martínez, Supervisor
-1996-1999, Instructor, Cellular Biology laboratory. INTA, Universidad de Chile, Santiago, Chile
-2000- March 2008, Assistant Professor. Cellular Biology Laboratory, INTA, Universidad de Chile, Santiago, Chile

Research's areas:

Biomedicine; Cancer: TGF-beta Signal transduction, Cell migration and invasion, extracellular matrix proteinases regulation.

Professional Appointments:

Participation in 14 Research's projects, 3 as a head researcher.

-41 international publications in peer reviewed journals
- More than 60 presentations in Chilean and international scientific meetings.
-Reviewer in international journals: European Journal of Cancer, FEBS letters and Life Sci.
-Reviewer of Chilean grants of FONDECYT (Chilean agency for projects), from 2006 to date.
- Director of Research office (2007), INTA, University of Chile

Honors:

1. "Zanlungo Foundation" 1989. Dpto. de Química Facultad de Ciencias. Universidad de Santiago de Chile. Chile.
2. Young research . "Abraham Stekel program" INTA, Universidad de Chile. 1992 – 1996, Chile
3. Mutis (AECI/ICI), 1997-1999, Biological Doctoral program, Autonoma University /IIB Madrid, Spain.
4. Andes Foundation (Chile), Short program fellowship. May-Jun 2000.IIB, Madrid Spain
5. Academic Stimulus", 2000, INTA, Universidad de Chile
6. " Foreign Doctors in Spain program" Ministerio de Educación, Cultura y Deportes (MECD), Spain. Centro de Investigaciones Biológicas, CSIC, Madrid, España. March 2003-August 2004
7. Fonds de la Recherche Scientifique-FNRS, Brussels . Belgium. Laboratory of Pharmacocinétique, Métabolisme, Nutrition e Toxicologie. Université Catholique de Louvain. Oct-Dec 2007.

Publications during Last Three Years:

1. Kocic J, Bugarski D, Santibanez JF. Smad3 is essential for TGF-Beta1-induced urokinase type plasminogen activator expression and migration in transformed keratinocytes. *European J. Cancer.* 2011, (submitted, under revision)
2. Santibáñez JF, Quintanilla M and Bernabéu C. TGF- /TGF- receptor system and its role in physiological and pathological conditions. Clinical Sci. 2011 (in press)
3. Pérez-Gómez E, del Castillo G, *Santibáñez JF*, López-Novoa JM, Bernabéu C, and M. Quintanilla. The role of the TGF-b co-receptor endoglin in cancer. *ScientificWorldJournal.* 2010; 10:2367-84.
4. Villar V, Kocic J, Bugarski D, Jovcic G, and Santibanez JF. SKIP is required for TGF-b1-induced epithelial mesenchymal transition and migration in transformed keratinocytes. *FEBS Lett.* 2010; 584(22):4586-92.
5. Santibáñez JF, Pérez-Gómez E, Fernández-L A, Carnero A, Malumbres M, Quintanilla M and Bernabéu C. The TGF-b co-receptor endoglin modulates the expression and transforming potential of H-Ras. *Carcinogenesis.* 2010; 31(12):2145-54.
6. Krsti M., Sovilj S., Grguri -Šipka S., Radosavljevi Evans I., Borozan S., Santibanez JF., Koci J. New ruthenium(II) complexes with N-alkylphenothiazines: Synthesis, structure, in vivo activity as free radical scavengers and in vitro cytotoxicity. *Eur J Med Chem.* 2010; 45(9):3669-76
7. Santibáñez JF, Koci J, Fabra A, Cano A, Quintanilla M. Rac1 modulates TGF-beta1-mediated epithelial cell plasticity and MMP9 production in transformed keratinocytes. *FEBS Lett.* 2010; 584(11):2305-10
8. Tobar N, Villar V, Santibanez JF. ROS-NFkappaBeta mediates TGF-beta1-induced expression of urokinase-type plasminogen activator, matrix metalloproteinase-9 and cell invasion. *Mol Cell Biochem.* 2010; 340(1-2):195-202
9. Krsti A, Santibanez JF, Oki I, Mojsilovi S, Koci J, Jov i G, Milenkovi P, Bugarski D. Combined effect of IL-17 and blockade of nitric oxide biosynthesis on haematopoiesis in mice. *Acta Physiol (Oxf).* 2010; 199: 31-41
10. Villar V, Koci J, Santibanez JF. Spred2 inhibits TGF-beta1-induced urokinase type plasminogen activator expression, cell motility and epithelial mesenchymal transition. *Int J Cancer.* 2010 Jul 1;127(1):77-85.
11. Echeverria C , Santibañez JF , Donoso-Tauda O , Escobar C and Ramirez-Tagle R. Structural Antitumoral Activity Relationships of Synthetic Chalcones. *Int. J. Mol. Sci.* 2009; 10(1), 221-231
12. Tobar N, Caceres M, Santibanez JF, Smith PC, Martinez J. RAC1 activity and intracellular ROS modulate the migratory potential of MCF-7 cells through a NADPH oxidase and NFkappaB-dependent mechanism. *Cancer Lett.* 2008; 267(1):125-32.
13. Santibanez JF, Blanco FJ, Garrido-Martin EM, Sanz-Rodriguez F, del Pozo MA, Bernabeu C. Caveolin-1 interacts and cooperates with the transforming growth factor-beta type I receptor ALK1 in endothelial caveolae. *Cardiovasc Res.* 2008;77(4):791-9.

In: Proteins Researcher Biographical Sketches … ISBN: 978-1-62100-777-7
Editors: H. Z. Wang and M. Tian

Chapter 47

BIOGRAPHICAL SKETCH

NAME:	TITLE:	DATE OF BIRTH:
Siracusano Luca	UNIVERSITY OF MESSINA	23\02\1948

EDUCATION:

Institution and Location	Degree	Year Conferred	Scientific Field
Department of Neuroscience, Psychiatric and Anesthesiological Sciences, University	MD Associated Professor		

CONTACT POINTS:

Address: viale Regina Margherita 28 98121 Messina Italy

RESEARCH AND PROFESSIONAL EXPERIENCE:

Intensive Care

Professional Appointments:

MD Associated Professor — MD Aggregated Professor

Publications during Last Three Years:

1) Siracusano L, Girasole V: Carbon monoxide and adiponectin in sepsis. Surgery. 2010 May;147(5):755.
2) Siracusano L, Girasole V: The genetics of malignant hyperthermia and related muscular syndromes. Anesth Analg. 2010 Apr 1;110(4):1241;.
3) Siracusano L, Girasole V: Propofol and cardioprotection against arrhythmias. Anesthesiology. 2009 Aug;111(2):447-8.
4) Siracusano L, Girasole V: Glucose and lipid metabolism in sepsis and endotoxemia. Acta Anaesthesiol Scand. 2009 Mar;53(3):413-4.
5) Siracusano L, Girasole V: Intercellular junctions in sepsis. Crit Care Med. 2008 Feb;36(2):659- 60; author reply 660-1.
6) Siracusano L, Girasole V: Sevoflurane and cardioprotection. Br *J* Anaesth. 2008 Feb;100(2):278; author reply 278-9.

In: Proteins Researcher Biographical Sketches ...
Editors: H. Z. Wang and M. Tian ISBN: 978-1-62100-777-7

Chapter 48

BIOGRAPHICAL SKETCH

NAME:	TITLE:	DATE OF BIRTH:
Yuji Takahashi	**DEPARTMENT OF REPRODUCTIVE BIOLOGY, NATIONAL CENTER FOR CHILD HEALTH AND DEVELOPMENT, JAPAN**	Mar 10, 1966

EDUCATION:

Institution and Location	Degree	Year Conferred	Scientific Field
Kyoto University	B.S.	1989	Animal Science
Kyoto University	M.S.	1991	Animal Science
University of Tokyo	Ph.D.	1997	Reproductive Biology

CONTACT POINTS:

Address: 2-10-1 Okura, Setagaya-ku, Tokyo 157-8535, Japan

RESEARCH AND PROFESSIONAL EXPERIENCE:

Staff Scientist, Laboratory of Pharmacology, Central Research Institute, Nissin Food Products Co., Ltd., Japan, 1991-1992

Postdoctoral Research fellow, Department of Cell Biology, University of Virginia Health System, School of Medicine, Charlottesville, VA 1997-2001

Postdoctoral Research Fellow, Laboratory of Animal Breeding, Graduate School of Agriculture and Life Science, University of Tokyo, Japan 2001-2003

Professional Appointments:

Professional Appointments: Scientific Director, National Center for Child Health and Development (2003 - 2011)

Publications during Last Three Years:

1. Toyoda M, Yamazaki-Inoue M, Itakura Y, Kuno A, Ogawa T, Yamada M, Akutsu H, Takahashi Y, Kanzaki S, Narimatsu H, Hirabayashi J, Umezawa A. (2011) Lectin microarray analysis of pluripotent and multipotent stem cells. Genes to Cell 16: 1-11.
2. Ito M, Miyado K, Nakagawa K, Muraki M, Imai M, Yamakawa N, Qin J, Hosoi Y, Saito H, Takahashi Y. (2010) Age-associated changes in the subcellular localization of phosphorylated p38 MAPK in human granulosa cells. Mol Hum Reprod 16: 926-937.
3. Imai M, Muraki M, Takamatsu K, Saito H and Takahashi Y. (2008) Spontaneous transformation of human granulosa cell tumours into an aggressive phenotype: a metastasis model cell line. BMC Cancer 8: 319.
4. Miyado K, Yoshida K, Yamagata K, Sakakibara K, Okabe M, Wang X, Miyamoto K, Akutsu H, Kondo T, Takahashi Y, Ban T, Itoh C, Toshimori K, Nakamura A, Ito M, Miyado M, Mekada E, and Umezawa A. (2008) The fusing ability of sperm is bestowed by CD9-containing vesicles released from eggs in mice. Proc Natl Acad Sci, USA 105: 12921-12926.
5. Horikawa T, Nakagawa K, Ohgi S, Kojima R, Nakashima A, Ito M, Takahashi Y, Saito H. (2008) The frequency of ovulation from the affected ovary decreases following laparoscopic cystectomy in infertile women with unilateral endometrioma during a natural cycle. J Assist Reprod Genet 25: 239-244.
6. Ito M, Muraki M, Takahashi Y, Imai M, Tsukui T, Yamakawa N, Nakagawa K, Ohgi S, Horikawa T, Iwasaki W, Iida A, Nishi Y, Yanase T, Nawata H, Miyado K, Kono T, Hosoi Y and Saito H. (2008) Glutathione S-transferase theta1 expressed in granulosa cells as a biomarker for oocyte quality in age-related infertility. Fertil Steril 90: 1026-1035.

In: Proteins Researcher Biographical Sketches ...
Editors: H. Z. Wang and M. Tian

ISBN: 978-1-62100-777-7

Chapter 49

BIOGRAPHICAL SKETCH

NAME:	TITLE:	DATE OF BIRTH:
Hiroki Tanabe	DEPARTMENT OF INTERNAL MEDICINE, GASTROENTEROLOGY AND HEMATOLOGY/ONCOLOGY, ASAHIKAWA MEDICAL UNIVERSITY	July 20, 1968

EDUCATION:

Institution and Location	Degree	Year Conferred	Scientific Field
Asahikawa Medical College	M.D.	1994	
Asahikawa Medical College	Ph.D.	1998	

CONTACT POINTS:

Address: 2-1-1-1 Midorigaoka-Higashi Asahikawa, Hokkaido 078-8510, Japan

RESEARCH AND PROFESSIONAL EXPERIENCE:

1999-2000: Medical doctor of Third Department of Internal Medicine, Asahikawa Medical College
2000-2003: a postdoctoral fellow in Department of Pathology, University of California, Irvine
2006-2007: Instructor of Department of Gastrointestinal Immunology and Regenerative Medicine, Asahikawa Medical College

2007-2009: Instructor of Department of Internal Medicine, Gastoenterology and hematological oncology, Asahikawa Medical College

2009-2011: Assistant director, Research and Development, Division Health policy Bureau, Ministry of Health, Labour and Welfare, Government of Japan

2011- present: Instructor of Department of Internal Medicine, Gastroenterology and Hematology oncology, Asahikawa Medical University

Professional Appointments:

Instructor of the department of Internal Medicine

Honors:

Special award in 12th Hamanako symposium, 2005

Publications during Last Three Years:

Ueno N, Fujiya M, Segawa S, Nata T, Moriichi K, Tanabe H, Mizukami Y, Kobayashi N, Ito K, Kohgo Y. Heat-killed body of lactobacillus brevis SBC8803 ameliorates intestinal injury in a murine model of colitis by enhancing the intestinal barrier function. Inflamm Bowel Dis. 2011 Jan 6.

Ueno N, Fujiya M, Moriichi K, Ikuta K, Nata T, Konno Y, Ishikawa C, Inaba Y, Ito T, Sato R, Okamoto K, Tanabe H, Maemoto A, Sato K, Watari J, Ashida T, Saitoh Y, Kohgo Y. Endosopic Autofluorescence Imaging is Useful for the Differential Diagnosis of Intestinal Lymphomas Resembling Lymphoid Heyperplasia. J Clin Gastroenterol. 2010 Oct 27.

Shanahan MT, Tanabe H, Ouellette AJ. Strain-specific polymorphisms in Paneth cell α-defensins of C57BL/6 mice and evidence of vestigial myeloid α-defensin pseudogenes. Infect Immun. 2011 Jan;79(1):459-73.

Watari J, Sakurai J, Morita T, Yamazaki T, Okugawa T, Toyoshima F, Tanaka J, Tomita T, Kim Y, Oshima T, Hori K, Moriichi K, Tanabe H, Fujiya M, Kohgo Y, Das KM, Matsumoto T, Miwa H. A case of early Barrett's adenocarcinoma repeatedly developing multiple metachronous lesions shortly after endoscopic therapy: an analysis for genetic and epigenetic alterations. Gastrointest Endosc. 2010 Dec;72(6):1303

Inaba Y, Ashida T, Ito T, Ishikawa C, Tanabe H, Maemoto A, Watari J, Ayabe T, Mizukami Y, Fujiya M, Kohgo Y. Expression of the antimicrobial peptide alpha-defensin/cryptdins in intestinal crypts decreases at the initial phase of intestinal inflammation in a model of inflammatory bowel disease, IL-10-deficient mice. Inflamm Bowel Dis. 2010 Sep;16(9):1488-95.

Sakamoto J, Fujiya M, Okamoto K, Nata T, Inaba Y, Moriichi K, Tanabe H, Mizukami Y, Watari J, Ashida T, Kohgo Y. Immunoprecipitation of nucleosomal DNA is a novel procedure to improve the sensitivity of serum screening for the p16 hypermethylation associated with colon cancer. Cancer Epidemiol. 2010 Apr;34(2):194-9.

Watari J, Moriichi K, Tanabe H, Sato R, Fujiya M, Miwa H, Das KM, Kohgo Y. Differences in genetic instability and cellular phenotype among Barrett's, cardiac, and gastric intestinal metaplasia in a Japanese population with Helicobacter pylori. Histopathology. 2009 Sep;55(3):261-9.

Ishikawa C, Tanabe H, Maemoto A, Ito T, Watari J, Kono T, Fujiya M, Ashida T, Ayabe T, Kohgo Y. Precursor Processing of Human Defensin-5 Is Essential to the Multiple Functions in vitro and in vivo. J Innate Immun 2009 2(1):66-76.

Moriichi K, Watari J, Das KM, Tanabe H, Fujiya M, Ashida T, Kohgo Y. Effects of Helicobacter pylori infection on genetic instability, the aberrant CpG island methylation status and the cellular phenotype in Barrett's esophagus in a Japanese population. Int J Cancer. 2009 Mar 15;124(6):1263-9.

Tanabe H, Sato T, Watari J, Maemoto A, Fijiya M, Kono T, Ashida T, Ayabe T, Kohgo Y. Functional Role of Metaplastic Paneth Cell Defensins in Helicobacter pylori-Infected Stomach Helicobacter 2008 Oct;13(5):370-379.

In: Proteins Researcher Biographical Sketches ...
Editors: H. Z. Wang and M. Tian
ISBN: 978-1-62100-777-7

Chapter 50

BIOGRAPHICAL SKETCH

NAME:	TITLE:	DATE OF BIRTH:
Gudmundur Thorgeirsson	**UNIVERSITY OF ICELAND AND LANDSPITALI-NATIONAL UNIVERSITY HOSPITAL, REYKJAVIK, ICELAND**	14 Mar 1946

EDUCATION:

Institution and Location	Degree	Year Conferred	Scientific Field
University of Iceland, Medical Faculty	MD	1973	
Case Western Reserve University, Cleveland, OHIO	PhD	1978	
University Hospitals of Cleveland, Resident and Fellow, Pathology, Internal Medicine and Cardiology:		1974-1982	

CONTACT POINTS:

Address: Klapparas 4, 110 Reykjavik, Iceland

Research: Endothelial cell biology, clinical cardiology, epidemiology and genetics of cardiovascular diseases.

Professional experience: Cardiologist, Landspitali-National University Hospital, since 1982

Currently: Professor and faculty chairman, Department of Medicine, Landspitali-National University Hospital; Dean, Faculty of Medicine, University of Iceland

Honors: FACP, FACC, FRCP (London)

Publications during Last Three Years:

1. B. Blaskó, R. Kolka, P. Thorbjornsdottir, S.T. Sigurdarson, G. Sigurdsson, Z. Rónai, M. Sasvári-Székely, S. Bödvarsson, G. Thorgeirsson, Z. Prohászka, M. Kovács, G. Fust, G.J. Arason: Low complement C4B gene copy number predicts short-term mortality after acute myocardial infarction. *Int Immunol*. 2008; 20:31-7.
2. A. Helgadottir, G. Thorleifsson, K.P. Magnusson, S. Grétarsdottir, V. Steinthorsdottir, A. Manulescu, G.T. Jones, G.J. Rinkel, J.D. Blankensteijn, A. Ronkainen, J.E. Jaaskelainen, Y. Kyo, G.M. Lenk, N. Sakalihasan, K. Kostulas. A. Gottsatter, A. Flex, H. Stefansson, T. Hansen, G. Andersen, S. Weinsheimer, K. Borch-Johnsen, T. Jorgensen, S.H. Shah, A.A. Ouyyumi, C.B. Granger, M.P. Reilly, H. Austin, A.I. Levey, V. Vaccarino, E. Palsdottir, G.B. Walters, T. Jonsdottir, S. Snorradottir, D. Magnusdottir, G. Gudmundsson, R.E. Ferrel, S. Sveinbjornsdottir, J. Hernesniemi, M. Niemela, R. Limet, K. Andersen, G. Sigurdsson, R. benediktsson, E.L. Verhoeven, J.A. Teijink, D.E. Grobbee, D.J. Rader, D.A. Collier, O. Pedersen, R. Pola, J. Hillert, B. Lindblad, E.M. Valdimarsson, H.B. Magnadottir, C. Wijmenga, G. Tromp, A.F. Baas, Y.M. Ruigrok, A.M. van Rij, H. Kuivaniemi, J.T. Powell, S.E. matthiasson, J.R. Gulcher, G. Thorgeirsson, A. Kong, U. Thorsteinsdottir, K. Stefansson: The same sequence variant on 9p21 associates with myocardial infarction, abdominal aortic aneurysm and intracranial aneurysm. *Nat Genet*. 2008; 40:217-24.
3. B.G. Libungan, K. Eyjolfsson, G. Thorgeirsson: Primary percutaneous coronary interventions in Iceland. *Laeknablaðið*. 2008;94:103-7.
4. B. Thors, H. Halldórsson, G. Jónsdóttir, G. Thorgeirsson: Mechanism of thrombin mediated eNOS phosphorylation in endothelial cells is dependent on ATP levels after stimulation.*Biochim Biophys Acta*. 2008;1783:1893-902.
5. S.Gretarsdottir, G.Thorleifsson, A.Manolescu, U.Styrkarsdottir, A.Helgadottir, A.Gschwendtner, K.Kostulas, G.Kuhlenbäumer, S.Bevan, T.Jonsdottir, H.Bjarnason, J.Saemundsdottir, S.Palsson, D.O.Arnar, H.Holm, G.Thorgeirsson, E.M.Valdimarsson, S.Sveinbjörnsdottir, C.Gieger, K.Berger, H.E.Wichmann, J.Hillert, H.Markus, J.R.Gulcher, E.B.Ringelstein, A.Kong, M.Dichgans, D.F.Gudbjartsson, U.Thorsteinsdottir, K.Stefansson: Risk variants for atrial fibrillation on chromosome 4q25 associate with ischemic stroke. *Ann Neurol* 2008; 64:402-9.
6. M.L. Scherer, T. Aspelund, S. Sigurdsson, R. Detrano, M. Garcia, G.F. Mitchell, L.J. Launer, G. Thorgeirsson, V. Gudnason, T.B. Harris. Abnormal T-wave axis is associated with coronary artery calcification in older adults. *Scand Cardiovasc J*. 2009;43:240-8.

7. Gudbjartsson DF, Holm H, Gretarsdottir S, Thorleifsson G, Walters GB, Thorgeirsson G, Gulcher J, Mathiesen EB, Njølstad I, Nyrnes A, Wilsgaard T, Hald EM, Hveem K, Stoltenberg C, Kucera G, Stubblefield T, Carter S, Roden D, Ng MC, Baum L, So WY, Wong KS, Chan JC, Gieger C, Wichmann HE, Gschwendtner A, Dichgans M, Kuhlenbäumer G, Berger K, Ringelstein EB, Bevan S, Markus HS, Kostulas K, Hillert J, Sveinbjörnsdóttir S, Valdimarsson EM, Løchen ML, Ma RC, Darbar D, Kong A, Arnar DO, Thorsteinsdottir U, Stefansson K.: A sequence variant in ZFHX3 on 16q22 associates with atrial fibrillation and ischemic stroke. *Nat Genet.* 2009;41:876-8.
8. Gudmundsson LS, Aspelund T, Scher AI, Thorgeirsson G, Johannsson M, Launer LJ, Gudnason V.: C-reactive protein in migraine sufferers similar to that of non-migraineurs: The Reykjavik Study. Cephalalgia. 2009 Apr 28.
9. Gudbjartsson DF, Bjornsdottir US, Halapi E, Helgadottir A, Sulem P, Jonsdottir GM, Thorleifsson G, Helgadottir H, Steinthorsdottir V, Stefansson H, Williams C, Hui J, Beilby J, Warrington NM, James A, Palmer LJ, Koppelman GH, Heinzmann A, Krueger M, Boezen HM, Wheatley A, Altmuller J, Shin HD, Uh ST, Cheong HS, Jonsdottir B, Gislason D, Park CS, Rasmussen LM, Porsbjerg C, Hansen JW, Backer V, Werge T, Janson C, Jönsson UB, Ng MC, Chan J, So WY, Ma R, Shah SH, Granger CB, Quyyumi AA, Levey AI, Vaccarino V, Reilly MP, Rader DJ, Williams MJ, van Rij AM, Jones GT, Trabetti E, Malerba G, Pignatti PF, Boner A, Pescollderungg L, Girelli D, Olivieri O, Martinelli N, Ludviksson BR, Ludviksdottir D, Eyjolfsson GI, Arnar D, Thorgeirsson G, Deichmann K, Thompson PJ, Wjst M, Hall IP, Postma DS, Gislason T, Gulcher J, Kong A, Jonsdottir I, Thorsteinsdottir U, Stefansson K.: Sequence variants affecting eosinophil numbers associate with asthma and myocardial infarction. *Nat Genet.* 2009;4:342-7.
10. Holm H, Gudbjartsson DF, Arnar DO, Thorleifsson G, Thorgeirsson G, Stefansdottir H, Gudjonsson SA, Jonasdottir A, Mathiesen EB, Njølstad I, Nyrnes A, Wilsgaard T, Hald EM, Hveem K, Stoltenberg C, Løchen ML, Kong A, Thorsteinsdottir U, Stefansson K: Several common variants modulate heart rate , PR interval and QRS duration. *Nat Genet.* 2010; 42:117-22.
11. Gudbjornsson B, Thorsteinsson SB, Sigvaldason H, Einarsdottir R, Johannsson M, Zoega H, Halldorsson M, Thorgeirsson G: Rofecoxib, but not celecoxib, increases the risk of thromboembolic cardiovascular events in young adults – a nationwide registry-based study. *Eur J Clin Pharmacol.* 2010; 66:619-25.
12. Vidal JS, Sigurdsson S, Jonsdottir MK, Eiriksdottir G, Thorgeirsson G, Kjartansson O, Garcia ME, van Buchem MA, Harris TB, Gudnason V, Launer LJ: Coronary artery calcium, brain function and structure: the AGES-Reykjavik Study. *Stroke.* 2010;41:891-7.
13. Gretarsdottir S, Baas AF, Thorleifsson G, Holm H, den Heijer M, de Vries JP, Kranendonk SE, Zeebregts CJ, van Sterkenburg SM, Geelkerken RH, van Rij AM, Williams MJ, Boll AP, Kostic JP, Jonasdottir A, Jonasdottir A, Walters GB, Masson G, Sulem P, Saemundsdottir J, Mouy M, Magnusson KP, Tromp G, Elmore JR, Sakalihasan N, Limet R, Defraigne JO, Ferrell RE, Ronkainen A, Ruigrok YM, Wijmenga C, Grobbee DE, Shah SH, Granger CB, Quyyumi AA, Vaccarino V, Patel RS, Zafari AM, Levey AI, Austin H, Girelli D, Pignatti PF, Olivieri O, Martinelli N, Malerba G, Trabetti E, Becker LC, Becker DM, Reilly MP, Rader DJ, Mueller T, Dieplinger B, Haltmayer M, Urbonavicius S, Lindblad B, Gottsäter A, Gaetani E, Pola R, Wells P, Rodger M, Forgie M, Langlois N, Corral J, Vicente V, Fontcuberta J, España F, Grarup N, Jørgensen T,

Witte DR, Hansen T, Pedersen O, Aben KK, de Graaf J, Holewijn S, Folkersen L, Franco-Cereceda A, Eriksson P, Collier DA, Stefansson H, Steinthorsdottir V, Rafnar T, Valdimarsson EM, Magnadottir HB, Sveinbjornsdottir S, Olafsson I, Magnusson MK, Palmason R, Haraldsdottir V, Andersen K, Onundarson PT, Thorgeirsson G, Kiemeney LA, Powell JT, Carey DJ, Kuivaniemi H, Lindholt JS, Jones GT, Kong A, Blankensteijn JD, Matthiasson SE, Thorsteinsdottir U, Stefansson K: Genome-wide association study identifies a sequence variant within the DAB2IP gene conferring susceptibility to abdominal aortic aeurysm. *Nat Genet*. 2010; 42:692-7.

14. Gudmundsson LS, Scher AI, Aspelund T, Eliasson JH, Johannsson M, Thorgeirsson G, Launer L, Gudnason V: Migraine with aura and risk of cardiovascular and all cause mortality in men and women: prospective cohort study. *BMJ*. 2010;341:c3966.
15. Assimes TL, Hólm H, Kathiresan S, Reilly MP, Thorleifsson G, Voight BF, Erdmann J, Willenborg C, Vaidya D, Xie C, Patterson CC, Morgan TM, Burnett MS, Li M, Hlatky MA, Knowles JW, Thompson JR, Absher D, Iribarren C, Go A, Fortmann SP, Sidney S, Risch N, Tang H, Myers RM, Berger K, Stoll M, Shah SH, Thorgeirsson G, Andersen K, Havulinna AS, Herrera JE, Faraday N, Kim Y, Kral BG, Mathias RA, Ruczinski I, Suktitipat B, Wilson AF, Yanek LR, Becker LC, Linsel-Nitschke P, Lieb W, König IR, Hengstenberg C, Fischer M, Stark K, Reinhard W, Winogradow J, Grassl M, Grosshennig A, Preuss M, Schreiber S, Wichmann HE, Meisinger C, Yee J, Friedlander Y, Do R, Meigs JB, Williams G, Nathan DM, MacRae CA, Qu L, Wilensky RL, Matthai WH Jr, Qasim AN, Hakonarson H, Pichard AD, Kent KM, Satler L, Lindsay JM, Waksman R, Knouff CW, Waterworth DM, Walker MC, Mooser VE, Marrugat J, Lucas G, Subirana I, Sala J, Ramos R, Martinelli N, Olivieri O, Trabetti E, Malerba G, Pignatti PF, Guiducci C, Mirel D, Parkin M, Hirschhorn JN, Asselta R, Duga S, Musunuru K, Daly MJ, Purcell S, Eifert S, Braund PS, Wright BJ, Balmforth AJ, Ball SG; Myocardial Infarction Genetics Consortium; Wellcome Trust Case Control Consortium; Cardiogenics, Ouwehand WH, Deloukas P, Scholz M, Cambien F, Huge A, Scheffold T, Salomaa V, Girelli D, Granger CB, Peltonen L, McKeown PP, Altshuler D, Melander O, Devaney JM, Epstein SE, Rader DJ, Elosua R, Engert JC, Anand SS, Hall AS, Ziegler A, O'Donnell CJ, Spertus JA, Siscovick D, Schwartz SM, Becker D, Thorsteinsdottir U, Stefansson K, Schunkert H, Samani NJ, Quertermous T: Lack of association between the Trp719Arg polymorphism in kinesin-like protein-6 and coronary artery disease in 19 case-control studies. *J Am Coll Cardiol*. 2010; 56:1552-63.
16. Thors B, Halldorsson, H, Thorgeirsson, G: eNOS activation mediated by AMPK after stimulation of endothelial cells with histamine or thrombin is dependent on LKB1, Biochimica et Biophysica Acta - Molecular Cell Research. 2011; 1813: 322-31.

In: Proteins Researcher Biographical Sketches … ISBN: 978-1-62100-777-7
Editors: H. Z. Wang and M. Tian

Chapter 51

BIOGRAPHICAL SKETCH

NAME:	TITLE:	DATE OF BIRTH:
Van Steendam Katleen	PHD AND PHARMD IN LABORATORY OF PHARMACEUTICAL BIOTECHNOLOGY, UNIVERSITY GHENT	05 April 1983

EDUCATION:

Institution and Location	Degree	Year Conferred	Scientific Field
Master in Pharmaceutical Sciences	PhD		Pharmaceutical Sciences

CONTACT POINTS:

Address: Harelbekestraat 72, 9000 Ghent, Belgium

RESEARCH AND PROFESSIONAL EXPERIENCE:

2006-2011: PhD student with a fellowship of the Fund of Scientific Research Flanders (Belgium) at the laboratory of Pharmaceutical Biotechnology, University Ghent

2011-present: Postdoctoral fellow at the laboratory of Pharmaceutical Biotechnology, University Ghent

Honors:

"Prix Pharmacien et Doctoresse Nedeljkovitch" 2006, Belgian Society for Pharmaceutical Sciences, Brussels, Belgium

Publications during Last Three Years:

Tilleman k., van steendam k., cantaert t., de keyser f., elewaut d., deforce d. Synovial detection and autoantibody reactivity of processed citrullinated isoforms of vimentin in inflammatory arthritides. Rheumatology, 2008, 47(5):597-604

Van steendam k., tilleman k., deceuleneer m., de keyser f., elewaut d., deforce d. Citrullinated vimentin as an important antigen in immune complexes from synovial fluid of rheumatoid arthritis patients with antibodies against citrullinated proteins arthritis research & therapy, 2010, 12:r132

Van steendam k., tilleman k., deforce d. The relevance of citrullinated vimentin in the production of antibodies against citrullinated proteins and the pathogenesis of rheumatoid arthritis *rheumatology 2011*;50(5):830-7

Van steendam k., van praet j., smith v., de bruyne g., mimori t., bonroy c., elewaut d., deforce d. And de keyser f. Specific antinuclear antibodies in systemic sclerosis patients with and without skin involvement: an extended methodological approach rheumatology 2011 feb 17, epub ehead of print

Deceuleneer m., de wit v., van steendam k., van nieuwerburgh f., tilleman k., deforce d. Modification of citrulline residues with 2,3-butanedione facilitates their detection by lc-ms rapid communications in Mass Spectrometry 2011 Jun 15;25(11):1536-42. doi: 10.1002/rcm.5015

In: Proteins Researcher Biographical Sketches ...
Editors: H. Z. Wang and M. Tian
ISBN: 978-1-62100-777-7

Chapter 52

BIOGRAPHICAL SKETCH

NAME:	TITLE:	DATE OF BIRTH:
Vishwanath Venketaraman	DR.	

Biography:

Upon completion of his doctorate in microbiology and immunology at the Tuberculosis Research Center in Chennai, India, Dr. Venketaraman was awarded a UNESCO fellowship to conduct post-doctoral research on tuberculosis at the University of Ferrara, Italy. Dr. Venketaraman continued his post-doctoral research at the University of Michigan Medical School and at the UMDNJ-New Jersey Medical School. Mycobacterial research has always fascinated Dr. Venketaraman. Specifically, Dr. Venketaraman is interested in characterizing the host immune defense mechanisms that are crucial for the control of *Mycobacterium tuberculosis*. Dr. Venketaraman has published more than 22 papers in peer-reviewed journals. Dr. Venketaraman is an Assistant Professor at the College of Osteopathic Medicine of Pacific, Western University of Health Sciences, CA. Dr. Venketaraman's laboratory is actively trying to elucidate the role of glutathione in enhancing the host immune cell functions to control of *Mycobacterium tuberculosis*, a novel and previously undescribed phenomenon. More than eight publications were generated from this body of work. Dr. Venketaraman's long term goal is to discover immunomodulatory agents that can be given an adjunct to chemotherapy for controlling *Mycobacterium tuberculosis* infections in both non-HIV and HIV-infected individuals.

In: Proteins Researcher Biographical Sketches … ISBN: 978-1-62100-777-7
Editors: H. Z. Wang and M. Tian

Chapter 53

BIOGRAPHICAL SKETCH

NAME:	TITLE:	DATE OF BIRTH:
Irina M. Vlasova	AFFILIATION: RUSSIA, MOSCOW, M.V. LOMONOSOV MOSCOW STATE UNIVERSITY, DEPARTMENT OF PHYSICS	

EDUCATION:

Institution and Location	Degree	Year Conferred	Scientific Field
M.V. Lomonosov Moscow State University, Department of Physics	Ph.D.		

CONTACT POINTS:

Address: Russia, 119991, Moscow, Leninskie Gori, M.V. Lomonosov Moscow State University, Department of Physics, Chair of General Physics

RESEARCH AND PROFESSIONAL EXPERIENCE:

1996-2002 M.Sc. Physics, chair Biophysics• , Department of Physics, Moscow State University
2002-2005 Post-graduated Student
2005 – Ph. D.
5/2005- 9/2005 – Junior Researcher
9/2005 – 3/2009 - Assistant Professor,

3/2009 – nowadays – Associated Professor, Department of Physics, Moscow State University
2010 -- nowadays - Chief of scientific group The spectroscopic investigations of blood components

Honors:

The winner of Moscow State University stipendiary for young scientific workers in 2006, 2007, 2009 years.

Publications during Last Three Years:

1. Vlasova I.M., Saletsky A.M. Investigation of denaturation of human serum albumin under action of cethyltrimethylammonium bromide by Raman spectroscopy. // Laser Physics, 2011, v. 21, â„– 1, p. 239-244.
2. Vlasova I.M., Bukharova E.M., Saletsky A.M. Rotational Diffusion of Fluorescein Family Markers in Solutions of Human Serum Albumin. // Russian Journal of Physical Chemistry A, 2011, v. 85, â„– 5, p. 874-878.
3. Vlasova I.M., Vlasov A.A., Saletsky A.M. Interaction of ionic detergent cethyltrimethylammonium bromide with human serum albumin at various values of pH: spectroscopic study. // Journal of Molecular Structure, 2010, v. 984. p. 332-338.
4. Vlasova I.M., Saletsky A.M. Raman spectroscopy in investigations of secondary structure of human serum albumin at binding of nanomarkers of fluorescein family. // Laser Physics, 2010, v. 20, â„– 9, p. 1844-1848.
5. Vlasova I.M., Saletsky A.M. Spectroscopic investigations of binding of three fluorescent nanomarkers to bionanomolecules of human serum albumin in dependence on pH. // Current Applied Physics, 2009, v. 9, â„– 5, p. 1027-1031.
6. Vlasova I.M., Saletsky A.M. Investigation of influence of different values of pH on mechanisms of binding of human serum albumin with markers of fluorescein family. // Journal of Molecular Structure, 2009, v. 936, p. 220-227.
7. Vlasova I.M., Saletsky A.M. Investigation of neuroprotective action of drug â€œSemaxâ€• at ischemic insult by Raman spectroscopy method by estimation of damage of low density lipoprotein of rat blood. // Laser Physics, 2009, v. 19, â„– 12, p. 2219-2223.
8. Vlasova I.M., Saletsky A.M. Raman spectroscopy in comparative investigations of mechanisms of binding of three molecular probes – fluorescein, eosin and erythrosin – to human serum albumin. // Laser Physics Letters, 2008, v. 5, â„– 11, p. 834-839.
9. Vlasova I.M., Saletsky A.M. Raman spectroscopy in investigations of mechanism of binding of human serum albumin to molecular probe fluorescein. // Laser Physics Letters, 2008, v. 5, â„– 5, p. 384-389.
10. Buravtcov D.E., Vlasova I.M., Saletsky A.M. Raman spectroscopy and fluorescence analysis in investigation of protective action of ischemic preconditioning at ischemic insult by estimation of damage of low density lipoprotein of blood. // Photomedicine and Laser Surgery, 2008, v. 26, â„– 3, p. 181-187.

In: Proteins Researcher Biographical Sketches … ISBN: 978-1-62100-777-7
Editors: H. Z. Wang and M. Tian

Chapter 54

BIOGRAPHICAL SKETCH

NAME:	TITLE:	DATE OF BIRTH:
Jiapu Zhang	THE UNIVERSITY OF BALLARAT	03/03/2011

EDUCATION:

Institution and Location	Degree	Year Conferred	Scientific Field
	PhD, MSc, MSc, BSc, CSIRO Postdoc		

CONTACT POINTS:

Address: Graduate School of Sciences, IT and Engineering, The University of Ballarat, Mount Helen, Ballarat, Victoria 3353, Australia, Phone: 61-4-23487360, Emails: jiapu_zhang@hotmail.com, j.zhang@ballarat.edu.au

RESEARCH AND PROFESSIONAL EXPERIENCE:

prion and Aβ

Professional Appointments:

Research Scientist of CSIRO and Research Fellow of the University of Melbourne and the University of Ballarat

Publications during Last Three Years:

Selected peer reviewed journal papers:

-Zhang JP and Sun J (2011) Optimal atomic-resolution structures of prion AGAAAAGA amyloid fibrils, J. of Theoretical Biology, DOI: 10.1016/j.jtbi.2011.02.012 (in press)

-Zhang JP (2011) Optimal molecular structures of prion AGAAAAGA palindrome amyloid fibrils formatted by simulated annealing, Journal of Molecular Modeling 17 (1): 173-179. PMID: 20411399: see Protein Crystallography Newsletter Volume 3, No. 1, January 2011, Crystallography Times http://www.rigakumsc.com/downloads/newsletter/LifeSciencesV03N01.html

-Zhang JP (2011), Comparison studies of the structural stability of rabbit prion protein with human and mouse prion proteins - . PMID: 20970434

-Zhang JP (2011), An effective simulated annealing refined replica exchange Markov chain Monte Carlo algorithm for the infectious disease model of H1N1 influenza pandemic, World Journal of Modelling and Simulation, 7(1): 29-39.

-Zhang JP (2010), Studies on the structural stability of rabbit prion probed by molecular dynamics simulations of its wild-type and mutants, J. of Theoretical Biology 264 (1) 119-122. PMID: 20109469

-Zhang JP (2010), Compare the replica exchange sampling strategy with the Metropolis-Hastings sampling strategy of Markov chain Monte Carlo simulations for the H1N1 influenza pandemic infectious disease model, International Journal of Mathematical Modeling, Simulation and Applications 3(1) 99-109.

-Zhang JP (2009), Studies on the structural stability of rabbit prion probed by molecular dynamics simulations, Journal of Biomolecular Structure and Dynamics 27 (2), 159-162. PMID: 19583441

-Bagirov AM, Rubinov AM, Zhang JP (2009), A multidimensional descent method for global optimization, Optimization 58 (5), 611-625.

In: Proteins Researcher Biographical Sketches ...
Editors: H. Z. Wang and M. Tian
ISBN: 978-1-62100-777-7

Chapter 55

BIOGRAPHICAL SKETCH

NAME:	TITLE:	DATE OF BIRTH:
Zoccola Marina	**CNR-ISMAC**	07/04/62

EDUCATION:

Institution and Location	Degree	Year Conferred	Scientific Field
	Degree in Agricultural Science		

CONTACT POINTS:

Address: C.so Pella 16, 13900, Biella, Italy

RESEARCH AND PROFESSIONAL EXPERIENCE:

Since 1989 she has been working as a researcher at the National Research Council, Institute for macromolecular Studies, in Biella. Her principal interests are in the study and characterization of biopolymers, mainly structural proteins (wool, fine animal fibres, silk, human hair) and they concern:

- production of keratin-based composite bioplastics
- production of polymer nanofibres filters for biomedical and filtering sectors
-melanin determination in human hair for the identification of subjects at high risk of developing skin tumours
-characterization of the dust produced in carding and combing wool mills and the evaluation of the respiratory risk of workers

-characterization of the Italian wools for the determination of the technical and operative conditions to valorise Italian wool fibres.

Publications during Last Three Years:

- M. Zoccola , A. Boschi, C. Arosio , R. Mossotti, R. Innocenti and G.Freddi, "Silk Grafting with Methacrylamide: a Near Infrared Spectroscopy Study", Journal of Applied Polymer Science, DOI#33154.
-M. Zoccola, A .Aluigi, C. Tonin Characterisation of keratin biomass from butchery and wool industry wastes, Journal of Molecular Structure, 938 (2009) 35-40.
- M. Zoccola, A. Aluigi, C. Vineis, F. Ferrero, M. Piacentino, C. Tonin, "Study on nanofibrous membranes made from keratin/fibroin blends", Biomacromolecules 9(10) (2008) 2819–2825.

PART 2 – RESEARCH SUMMARIES IN PROTEINS

In: Proteins Researcher Biographical Sketches ...
Editors: H. Z. Wang and M. Tian
ISBN: 978-1-62100-777-7

Chapter 56

SMALL STRESS PROTEINS AND HUMAN DISEASES

Stéphanie Simon and André-Patrick Arrigo
Claude Bernard University, Villeurbanne, France

RESEARCH SUMMARY

Essential for the defense against all kinds of stress, small stress proteins also play major roles in the physiology, development and differentiation of human cells. In spite of their beneficial roles, small stress proteins can also be deleterious for the organism, notably in some cancer pathologies where they can, through their anti-apoptotic properties, favor tumor development and interfere with the cytotoxic activity of anti-cancer drugs. This book proposes a focus on the recent knowledge in the field of the expression and putative role of small stress proteins in human health and diseases.

In: Proteins Researcher Biographical Sketches ...
Editors: H. Z. Wang and M. Tian
ISBN: 978-1-62100-777-7

Chapter 57

ELUCIDATING THE GROWTH REGULATION OF BRASSINOSTEROIDS IN MUNG BEAN EPICOTYLS USING A PROTEOMICS APPROACH

Bin Huang[1], Kuo-Chin Ni[2], Shu-Ling Chen[3], Hsueh-Fen Juan*[2,4,5] and Yih-Ming Chen*[1,4]

[1] Institute of Plant Biology, National Taiwan University, Taipei, Taiwan
[2] Institute of Molecular and Cellular Biology, National Taiwan University, Taipei, Taiwan
[3] Department Biotechnology, Ming Chuan University, Taoyuan, Taiwan
[4] Department of Life Science, National Taiwan University, Taipei, Taiwan
[5] Center for Systems Biology and Bioinformatics, National Taiwan University, Taipei, Taiwan

RESEARCH SUMMARY

Brassinosteroids (BRs), structurally similar to animal steroid hormones, are ubiquitously distributed through out the plant kingdom with significant growth-promoting activity. Even though the BR-regulated transcriptome has been revealed, the proteomics data are still rare. Mung bean seedlings treated with 24-epibrassinolide (EBL), a kind of BRs, exhibited significant epicotyl elongation. Here, we present a proteomics study of EBL-treated mung bean epicotyls. Many researches showed that *de novo* sequencing is very useful to identify the proteins in non-model organisms. We successfully used *de novo* sequencing to identify 12 differentially expressed proteins in mung bean treated with or without EBL. These proteins are mostly involved in cell growth, anti-stresses, and cell metabolism including respiration

---

* Corresponding author: Professor Hsueh-Fen Juan, Department of Life Science and Institute of Molecular and Cellular Biology, National Taiwan University, Taipei, Taiwan, No 1, Sec. 4, Roosevelt Road, Taipei, 106 Taiwan; E-mail: *yukijuan@ntu.edu.tw*; Tel: +886-2-33664536; Fax: +886-2-23673374

* Corresponding author: Professor Yih-Ming Chen, Department of Life Science and Institute of Plant Biology, National Taiwan University, Taipei, Taiwan, No 1, Sec. 4, Roosevelt Road, Taipei, 106 Taiwan; E-mail: *yihmingc@ntu.edu.tw*; Tel: +886-2-33662524; Fax: +886-2-83695080

and photosynthesis. In the physiological assay, the retarded growth caused by inhibitors of methionine synthase and enolase could be recovered by exogenous treatment of EBL. Based on the identified proteins, we proposed a possible mechanism of BR-promoted cell elongation. This study confirms that proteomic analysis is a potent tool for analyzing the protein expression related to the effect of steroid hormones on epicotyl growth. Our approach opens a way of defining hormonal targets for plant growth. This not only elucidates a molecular mechanism of BR regulation but may also be extended for plant breeding.

In: Proteins Researcher Biographical Sketches ...
Editors: H. Z. Wang and M. Tian

ISBN: 978-1-62100-777-7

Chapter 58

ASCERTAINING THE LONG-TERM EFFECTS OF ACUTE SYSTEMIC NEONATAL PROTEIN SYNTHESIS INHIBITION ON COGNITION AND BEHAVIOR IN SPRAGUE-DAWLEY RATS (*RATTUS NORVEGICUS*)

Robert W. Flint, Jr., Heather Joppich, Christina L. Marino, Leslie A. Sandusky, Sarah Valentine and Jonathan E. Hill
The College of Saint Rose, Albany, NY, US

RESEARCH SUMMARY

The early postnatal period represents a time of significant neural development. Neonatal exposure to stressors, environmental toxins, and many different pharmacological agents may have a significant impact on subsequent behavior and cognition in adulthood. The purpose of this study was to examine the effects of early administration of the protein synthesis inhibitor, cycloheximide, on a battery of behavioral and cognitive tests in adulthood. Twelve-day-old Sprague-Dawley rats were administered saline or .5 mg/kg of cycloheximide subcutaneously. Upon reaching adulthood, a battery of tests were administered in order to examine anxiety, locomotor activity, environmental habituation, spatial working memory, object recognition memory, long-term spatial memory, and conditioned taste aversion. Most tests failed to reveal any differences between postnatally exposed cycloheximide and saline animals. However, during the water maze probe test, cycloheximide exposed animals reached the original platform location significantly faster than saline controls. For the conditioned taste aversion test, during the initial exposure to a 10% sucrose solution (prior to the induction of the taste aversion) the cycloheximide animals consumed significantly more than saline controls, indicating an attenuation of neophobia. Although acquisition of a conditioned taste aversion was comparable in neonatally exposed cycloheximide and saline animals, there was a significant resistance to extinction by cycloheximide animals. These results suggest that animals exposed to cycloheximide on postnatal day 12 displayed enhanced long-term spatial memory, reduced neophobia, and enhanced memory for taste aversion conditioning as indicated by a resistance to extinction.

In: Proteins Researcher Biographical Sketches ...
Editors: H. Z. Wang and M. Tian

ISBN: 978-1-62100-777-7

Chapter 59

Proteomic Profiling of Rat Brain Discharged by Ultrasound Associated with High Frequency Electro-Magnetic Field

***Huo-Yen Hsiao*[1] *and Ing-Feng Chang**[*2, 3, 4]

[1] Fong Yuang City, Taiwan ROC
[2] Department of Biochemistry and Molecular Biology, U of Nevada, Reno, US
[3] Institute of Plant Biology, National Taiwan University, Taipei, Taiwan ROC
[4] Department of Life Science, National Taiwan University, Taipei, Taiwan ROC

Research Summary

Beta-amyloid deposition was discovered in the Ultrasound Associated with High Frequency Electro-Magnetic Field (UAHFEMF)-treated rat brain (Chang and Hsiao, 2005, Current Alzheimer Research 2: 559-569). This lead the author to hypothesize the characteristics of generalized pathological changes of Alzheimer disease might be modulated by the biological functions of UAHFEMF discharge energy. A liquid chromatography-based proteomic analysis was performed and identified 16 rat brain proteins that were in the UAHFEMF-treated brain tissues but not in the normal tissues. These include 14-3-3 zeta, 14-3-3 epsilon protein, and antioxidant proteins such as Cu/Zn superoxide dismutase, and aldehyde dehydrogenase that can be involved in the oxidative stress response. In addition, a human RhoGAP homolog (Genbank accession # P84107) previous identified was consistently found to be induced in the UAHFEMF-treated rat brain. Many of the identified proteins are post-translational modified, *i.e* N-terminus acetylation of eight proteins, oxidation of 14-3-3 epsilon, and phosphorylation of feretin. In this study we provide proteomic evidence of the UAHFEMF effect on rat brain. The identified proteins can be important in Alzheimer's disease pathogenesis.

---

[*] Correspondence author: Ing-Feng Chang; Institute of Plant Biology, National Taiwan University; Taipei, Taiwan; Tel: 011-886-2-33662534; Fax: 011-886-2-23918940; E-mail: ifchang@ntu.edu.tw

In: Proteins Researcher Biographical Sketches ...
Editors: H. Z. Wang and M. Tian

ISBN: 978-1-62100-777-7

Chapter 60

ANALYTICAL APPROACHES TO COMPARATIVE MODELING OF PROTEIN STRUCTURES*

Aleksandar Poleksic
Department of Computer Science, University of Northern Iowa, Cedar Falls IA, US

RESEARCH SUMMARY

Comparative modeling builds a three-dimensional structure of a target protein based on one or several related proteins of known structure. The underlying assumption is that the sequence homology between two proteins implies structural similarity, which in turn implies a similarity in their function. Most algorithms for aligning and scoring pairs of proteins are based on the dynamic programming techniques in conjunction with the residue-residue mutation scores. However, the pair-wise alignment techniques that use amino-acid substitution matrices and residue-type information are often unable to recognize the relationships between proteins that are known to be homologous. Profile-profile methods utilize sequence information contained in multiple sequence alignments corresponding to both target and the template family of proteins. By taking advantage of the distribution of amino-acid counts in the columns of the alignment profiles, these methods can often recognize the similarity between remotely related sequences, as evidenced by large scale benchmarking experiments such as LiveBench, CAFASP, and CASP. We give an overview of the existing profile-profile techniques, concentrating on the key aspects of the alignment algorithms. A novel approach to scoring pairs of profiles is presented that focuses on the alignment scoring function and the estimation of the statistical parameters for the alignment score statistics.

* Reviewed by Natarajan Kannan, Ph.D., Howard Hughes Medical Institute, Department of Chemistry and Biochemistry, University of California San Diego, La Jolla, CA 92093.

In: Proteins Researcher Biographical Sketches …
Editors: H. Z. Wang and M. Tian
ISBN: 978-1-62100-777-7

Chapter 61

EVALUATION OF A SYNTHETIC BIOMIMETIC LIGAND FOR THE PURIFICATION OF THERAPEUTIC PROTEINS

Dimitris Platis and Nikolaos E. Labrou*
Laboratory of Enzyme Technology, Department of Agricultural Biotechnology,
Agricultural University of Athens, Iera Odos, GR Athens, Greece

RESEARCH SUMMARY

The successful commercialization of high-purity proteins and enzymes parallels the development and improvement of downstream processing technology. Biomimetic ligands find wide application in protein separation and purification, being the most promising affinity ligands of large-scale potential. In the present paper we evaluate the ability of the biomimetic ligand 4-amino-phenyl-oxanilic acid coupled to Sepharose CL-6B via 1,3,5-trichloro-2,4,6-triazine to bind and purify human monoclonal anti-HIV antibody 2F5 (mAb 2F5) from spiked maize extracts and recombinant influenza virus neuraminidase expressed in insect cells. Under selected conditions the affinity adsorbent exhibited high selectivity and purifying ability for both proteins.

* Corresponding author: E-mail: lambrou@aua.gr, Tel./fax +30 210 5294308

In: Proteins Researcher Biographical Sketches ...
Editors: H. Z. Wang and M. Tian
ISBN: 978-1-62100-777-7

Chapter 62

BIOINFORMATICS OF PROTEOMIC ANALYSIS OF RAT BRAIN DISCHARGED BY ULTRASOUND ASSOCIATED WITH HIGH FREQUENCY ELECTRO-MAGNETIC FIELD REVEALS POST-TRANSLATIONAL MODIFICATIONS

Ing-Feng Chang [1, 2, *] ***and Huo-Yen Hsiao*** [3]
[1]Institute of Plant Biology, National Taiwan University, Taipei, Taiwan ROC
[2]Department of Life Science, National Taiwan University, Taipei, Taiwan ROC
[3]Fong Yuang City, Taiwan ROC

RESEARCH SUMMARY

In the Ultrasound Associated with High Frequency Electro-Magnetic Field (UAHFEMF)-treated rat brain, deposition of beta-amyloid was discovered (Chang and Hsiao, 2005, Current Alzheimer Research 2: 559-569). The author hypothesizes the characteristics of generalized pathological changes of Alzheimer's disease might be modulated by the biological functions of UAHFEMF discharge energy. An in-depth bioinformatics analyses of the raw data of liquid chromatography-based proteomic analysis in a previous study identified 19 rat brain proteins in the UAHFEMF-treated brain tissues but not in the normal tissues of which 6 were significantly identified. These proteins include 14-3-3 zeta and 14-3-3 epsilon protein that can be involved in Alzheimer's disease pathogenesis. In particular, post-translational modifications such as N-terminus acetylation of nine proteins, oxidation of 14-3-3 epsilon, O-GlcNac modification of plasminogen, and phosphorylation of feretin were identified. In particular, the phosphorylation site of ferretin appeared to be a novel site which was never reported. These post-translational modifications can possibly be important in Alzheimer's disease pathogenesis.

* Corresponding author: Tel: 011-886-2-33662534, Fax: 011-886-2-23918940, E-mail: ifchang@ntu.edu.tw

In: Proteins Researcher Biographical Sketches ... ISBN: 978-1-62100-777-7
Editors: H. Z. Wang and M. Tian

Chapter 63

FOOD RESTRICTION EFFECTS ON PLASMA AMINO ACIDS, MYOFIBRILLAR PROTEIN, PLASMA AND MUSCLE FREE AND STERIFIED FATTY ACIDS ON MONOGASTRIC AND RUMINANTS: A REVIEW

André Martinho de Almeida*[*], *Sofia van Harten* and *Luís Alfaro Cardoso
Instituto De Investigação Científica Tropical – Tropical Research Institute, CVZ, Lisboa, Portugal

RESEARCH SUMMARY

Undernutrition is a major setback of animal production in the tropics. Progressive weight loss is characterized by a mobilization of body depots, especially fat and to a lesser extent protein. Muscle and therefore protein breaks down in order to supply amino acid as an energy source, generating a vast array of responses in serum free amino acid concentrations, myofibrillar protein (with special reference to Myosin Heavy Chains, Actin, Protein C, α Actinin and Tropomyosin + Troponin T) profiles and free fatty acids and fatty acids incorporated in triacylglycerol as a consequence of several factors: animal species and the severity of food restriction. The objective of this work is to present a review regarding these very important aspects of undernutrition. Several domestic animals species, both ruminant and monogastric will be considered with significance to both laboratory and farm animals. Special relevance will be given to works conducted at our group particularly on two experiments, first with laboratory rats and in a latter point *Boer goat* bucks in an applied prospective.

*Corresponding author: Tropical Research Institute, FMV, Rua Prof. Cid dos Santos, 1300-477 Lisboa, Portugal.

In: Proteins Researcher Biographical Sketches ... ISBN: 978-1-62100-777-7
Editors: H. Z. Wang and M. Tian

Chapter 64

Analytical Methods of the Determination of Arginine Amino Acid

R. M. Callejon, C. Ubeda, A. M. Troncoso and M. L. Morales
Área de Nutrición y Bromatología, Facultad de Farmacia,
Universidad de Sevilla, Sevilla, Spain

Research Summary

L-Arginine is one of the most versatile amino acids from the metabolic and physiological point of view. Known funtions for arginine include: substrate for protein and biologically active peptide synthesis, intermediate in ammonia detoxification, hormone liberation, and poliamine and creatine biosynthesis. All these funtions have been recently increased with the discovery of its role as the substrate for the synthesis of nitric oxide, a multifuntional effector involved in vasodilation, neurotransmission, and inmune responses. From the nutritional point of view, L-arginine must be considered as a “semiessential” or “conditionally essential” amino acid in mammals, depending on the developmental stage and health status of the individual. Therefore, the endogenous synthesis and dietary supply both contribute to satisfy arginine needs. In man, biosynthesis seems to be high enough to fulfil normal physiological arginine demand consumption of the amino acid. Such a situation suggests a potential therapeutic use for dietary arginine administration that is just starting to be analyzed. Hence, there is a need for the accurate and reliable quantification of arginine in food and various biological fluids. This current chapter gives a comprehensive overview of the techniques and methodologies currently available for the measurement of arginine in different matrixes. Arginine, as well as the rest of the amino acids, usually needs to be derivatized to make it more detectable. Several derivatizing reagents have been employed for the determination of arginine and each has its advantages and disadvantages.

In: Proteins Researcher Biographical Sketches ...
Editors: H. Z. Wang and M. Tian
ISBN: 978-1-62100-777-7

Chapter 65

ALTERNATIVE METABOLIC PATHWAYS OF ARGININE AND THEIR PATHOPHYSIOLOGICAL ROLES

András Hrabák and Zoltán Kukor*
Department of Medical Chemistry, Molecular Biology and Pathobiochemistry,
Semmelweis University, Budapest, Tűzoltó u., Hungary

RESEARCH SUMMARY

Arginine is not only a protein constituent but it is metabolized through various alternative pathways in mammalian cells. Our review is focused on the two most important alternative pathways: the nitric oxide (NO) synthesis and the arginase reaction. In cells where both pathways are active, their regulation is generally reciprocal due to the common substrate. This reciprocal regulation was described at the level of both enzyme activities and expressions. Cross inhibition between arginase and NO synthase metabolites can be observed and different cytokines act differently on the expression of both enzymes as well. Arginase and NO synthase isoenzymes are involved in various important physiological processes such as urea cycle, vasodilation, immune defense against certain invaders and neurotransmission. However, the overexpression of these enzymes and overproduction of NO may contribute to the onset of various pathophysiological processes. NO overproduction may be responsible for neurotoxicity, septic shock, type 1 diabetes mellitus or various inflammatory diseases in numerous organs. In addition, inflammatory responses involving the inducible NO synthase may also contribute to the etiology of metabolic syndrome, while impaired NO production by the endothelial NO synthase may be in the background of cardiovascular disorders or preeclampsia. On the other part, the involvement of high arginase expression was observed in cardiovascular disorders and several pulmonary diseases, as pulmonary hypertension, silicosis and asthma. In conclusion, alternative arginine metabolic pathways and their regulation are important factors both in the maintenance of healthy state and in the pathogenesis of various diseases.

---

* Postal Address: H-1444 POB 260, Hungary

In: Proteins Researcher Biographical Sketches ... ISBN: 978-1-62100-777-7
Editors: H. Z. Wang and M. Tian

Chapter 66

FREE AMINO ACID ANALYSIS IN NATURAL MATRICES

Graciliana Lopes, Patrícia Valentão and Paula B. Andrade*
REQUIMTE/Department of Pharmacognosy, Faculty of Pharmacy,
Porto University, R. Aníbal Cunha, Porto, Portugal

RESEARCH SUMMARY

Amino acids constitute a class of biologically active compounds found either in the free form or as linear chains in peptides and proteins. In addition to their primary function as protein components, they have several biological roles, being important as neurotransmitters, hormones, precursors of complex nitrogen containing molecules and as metabolic intermediates.

Amino acids are molecules containing an amine group, a carboxylic acid group and a side chain that varies between different amino acids. There are 20 amino acids commonly found in proteins, which are classified depending on the polarity of the side chain. According to this criterion they can be non-polar and neutral, polar and neutral, acidic or basic. In plants they are also involved in secondary metabolism, namely in the biosynthesis of phenolic compounds, glucosinolates, cyanogenic heterosides and alkaloids.

The amino acids composition is a reliable indicator of the nutritional value of matrices used for human consumption, but also a useful tool in natural products authenticity. In this work we provide an overview on the application of HPLC-UV-vis and GC-FID analysis on the determination of the amino acids profile of several natural matrices: wild edible mushroom species, *Brassica oleracea* var. *costata*, *Catharanthus roseus*, *Cydonia oblonga* and red wine inoculated with different *Dekkera bruxellensis* strains. The influence of different factors, such as the collection date, geographical origin and vegetal tissue, in the amino acids composition of the samples are also discussed.

In: Proteins Researcher Biographical Sketches ...
Editors: H. Z. Wang and M. Tian
ISBN: 978-1-62100-777-7

Chapter 67

DISCOVERY OF ARGININOSUCCINATE SYNTHETASE AND ARGININOSUCCINATE LYASE

Olivier Levillain*

Université Claude Bernard Lyon I, Physiologie Intégrative, Cellulaire et Moléculaire CNRS, Bâtiment. R. Dubois, Villeurbanne Cedex, France

RESEARCH SUMMARY

This review summarizes the impressive work performed by several teams of research to discover the biosynthesis of argininc in mammalian liver and kidney and the different steps which led to identify the metabolites involved in this reaction. Successive purification steps allowed the isolation and characterization of two enzymatic proteins, namely argininosuccinate synthetase and argininosuccinate lyase. New approaches including molecular biology gave insights into the molecular characteristics of ASS and ASL genes, mRNAs, and proteins. Furthermore, new techniques such as Northern blot, Western blot, and immunocytology became excellent tools to analyze the expression on ASS and ASL genes in the different organs of several mammalian species. ASS and ASL are homotetramers with subunits of 46 and 51 kDa, respectively. ASS and ASL are generally co-localized in the same cell type and are widely distributed in various organs.

* Email: Olivier.Levillain@univ-lyon1.fr
Fax 33-04-72-43-11-72
Phone 33-04-72-43-11-73

In: Proteins Researcher Biographical Sketches ... ISBN: 978-1-62100-777-7
Editors: H. Z. Wang and M. Tian

Chapter 68

EXPRESSION AND LOCALIZATION OF ARGININOSUCCINATE SYNTHETASE AND ARGININOSUCCINATE LYASE IN THE FEMALE AND MALE RAT KIDNEYS

Olivier Levillain,[1] and Heinrich Wiesinger*[2]
[1] Université Lyon I, Lyon, France,
Laboratoire de Physiologie Intégrative, Cellulaire et Moléculaire,
Bâtiment. R. Dubois, Villeurbanne Cedex, France
[2] Physiologisch-Chemisches Institut der Universität, Tuebingen, Germany

RESEARCH SUMMARY

The enzymes argininosuccinate synthetase (ASS) and argininosuccinate lyase (ASL) convert L-citrulline into L-arginine. In mammalian kidneys, L-arginine is essentially produced in the proximal convoluted tubules (PCT). Almost all the studies performed on this renal metabolic pathway were restricted to male animals. Our experiments were first conducted to determine whether female rat kidneys express ASS and ASL genes and the results were compared to those of the males. The expression of ASS and ASL was analyzed by Western blot analyses in the different zones dissected from rat kidneys. ASS protein was localized by immunofluorescence. Finally, we tested whether sex influences the renal level of ASS and ASL proteins, as determined by Western blot analyses. Our results reveal that high levels of ASS and ASL proteins were detected in female as well as in male rat kidneys. The relative abundance of ASS and ASL proteins was higher in the cortex (superficial > deep) than in the outer stripe of the outer medulla. Immunolocalization studies clearly showed that ASS expression was restricted to the proximal tubules. PCT exhibited the highest level of ASS compared with the proximal straight tubules (PST). The levels of ASS and ASL proteins were similar in female and male rat kidneys. In conclusion, female rat kidneys express the two enzymes involved in the production of the conditionally essential amino acid L-arginine. No sexual dimorphism in ASS and ASL expression was found in the rat kidney.

In: Proteins Researcher Biographical Sketches ...
Editors: H. Z. Wang and M. Tian

ISBN: 978-1-62100-777-7

Chapter 69

CHEMICAL STRUCTURE AND TOXICITY IN ARGININE-BASED SURFACTANTS

Aurora Pinazo[1*], Lourdes Pérez[1], María Rosa Infante[1], María Pilar Vinardell[2], Montse Mitjans[2], María Carmen Morán[2] and Verónica Martínez[2]

[1] Institut de Química Avançada de Catalunya, CSIC, Barcelona, Spain
[2] Facultat de Farmàcia, UB, Barcelona, Spain

RESEARCH SUMMARY

Surfactants are one of the most representative chemical products which are consumed in large quantities every day on a worldwide scale. The use of surfactants in everyday life is almost unavoidable. The development of less irritant, less toxic, consumer-friendly surfactants or surfactant systems is, therefore, of general interest. During the last 20 years, our group has been developing new biocompatible surfactants derived from amino acids. Among them, arginine derivative surfactants constitute a novel class that can be regarded as an alternative to conventional cationic surfactants due to their multifuncional properties and the renewable source of raw materials used during the synthesis process. These characteristics make them candidates of choice as additives *in* pharmaceutical, food and cosmetic formulations. Evaluation of the irritant potential in *vivo*, of new products or ingredients, for pharmaceutical use, is required by law in most EU countries, prior to human exposure. However, due to increasing concern over animal use and in lights of its potential ban in the near future, alongside with the obvious ethical implications of using directly human subjects, in vitro alternative methods should now be encouraged. This review reports on the relationship between the structure and toxicity evaluated by *in vitro* methods of a series of arginine-based surfactants including surfactants with one single chain, gemini surfactants, and surfactants with glycerolipid-like structure.

* E-mail: apgste@cid.csic.es

In: Proteins Researcher Biographical Sketches ...
Editors: H. Z. Wang and M. Tian
ISBN: 978-1-62100-777-7

Chapter 70

ARGININE: PHYSICO-CHEMICAL PROPERTIES, INTERACTIONS WITH ION-EXCHANGE MEMBRANES, RECOVERY AND CONCENTRATION BY ELECTRODIALYSIS

T. Eliseeva, E. Krisilova, G. Oros and V. Selemenev*
Voronezh State University, Universitetskaya, Voronezh, Russia

RESEARCH SUMMARY

L-Arginine (2-amino-5-guanidinpentanoic acid) is a product of great importance in medicine, food and pharmaceutical industry. L-Arginine is a basic, genetically coded α-amino acid, one of the twenty most common natural amino acids. It is essential for human ("semi-essential"). Arginine plays a significant role in cell division, healing of wounds, removing ammonia from a body, immune function, and release of hormones.

This amino acid is used to treat cardiovascular disorders as a precursor for the synthesis of nitric oxide (NO) such as heart failure, intermittent claudication, impotence, sexual disfunction, and interstitial cystitis.

The benefits and functions attributed to oral supplementation of L-arginine include:

- Reduces healing time of injuries (particularly, bones)
- Quickens repair time of damaged tissue
- Helps decrease blood pressure.

Arginine is the most basic amino acid (pI=10.76). Side chain of arginine consists of a 3-carbon aliphatic straight chain, the distal end of which is capped by a complex guanidinium group. Guanidinium group (pK_a 12.48) is positively charged in neutral, acidic and even most basic environments, and thus imparts basic chemical properties to arginine. Because of the

* Tel.: +74732208932; Fax: +74732208755;
E-mail: elena.vsu@mail.ru, tatyanaeliseeva@yandex.ru

conjugation between the double bond and the nitrogen lone pairs, the positive charge is delocalized, enabling the formation of multiple H-bonds.

As other amino acids, arginine is a typical ampholyte and exists in a solution in different ionic forms: cations, bipolar ions and anions. At the pH value, equal to isoelectric point, amino acids exist preferably in the form of bipolar ions in individual solution and even in solid state. In acidic solution bipolar ions recharge into cations:

$$^{+}H_3N\text{-}RCH\text{-}COO^{-} + H_3O^{+} \Leftrightarrow {}^{+}H_3N\text{-}RCH\text{-}COOH + H_2O \quad (1)$$

bipolar ion cation

At high pH value they can exist in negatively charged form:

$$^{+}H_3N\text{-}RCH\text{-}COO^{-} + OH^{-} \Leftrightarrow NH_2\text{-}RCH\text{-}COO^{-} + H_2O \quad (2)$$

bipolar ion anion

The aim of this article is to consider the processes of sorption, hydration and mass transfer in the system arginine-ion-exchange membrane-water as well as to show the possibilities of electrodialysis in separation of basic amino acid and concentration of its solutions.

In: Proteins Researcher Biographical Sketches ...
Editors: H. Z. Wang and M. Tian

ISBN: 978-1-62100-777-7

Chapter 71

CENTRAL FUNCTIONS OF L-ARGININE AND ITS METABOLITES FOR STRESS BEHAVIOR

***Shozo Tomonaga*[1], *D. Michael Denbow*[2] *and Mitsuhiro Furuse*[3*]**

[1] Laboratory of Advanced Animal and Marine Bioresources, Faculty of Agriculture, Kyushu University, Fukuoka, Japan
[2] Department of Animal and Poultry Sciences, Virginia Polytechnic Institute and State University, Blacksburg, VA, US
[3] Labotratory of Regulation in Metabolism and Behavior, Faculty of Agriculture, Kyushu University, Fukuoka, Japan

RESEARCH SUMMARY

L-Arginine is an essential amino acid for birds, carnivores and young mammals and a conditionally essential amino acid for adults. L-Arginine can be catabolized by four sets of enzymes in mammalian cells, resulting in the production of urea, L-ornithine, L-proline, L-glutamate, polyamines, nitric oxide, creatine, agmatine, etc.. Unlike mammals, birds lack carbamyl phosphate synthetase, one of the urea cycle enzymes necessary for the synthesis of L-citrulline from L-ornithine in the liver and kidney. Therefore, it is impossible to synthesize L-arginine in birds, and L-arginine is classified as an essential amino acid for birds. In this chapter, we introduce recent studies about central functions of L-arginine and its metabolites for stress behavior. In particular, the functions in avian species are focused upon. In neonatal chicks, centrally injected L-arginine induces sedative and hypnotic effects under social separation stress. Among L-arginine metabolites, L-ornithine, L-proline and L-glutamate would especially contribute to these effects.

* Correspondence to: Mitsuhiro Furuse
Laboratory of Regulation in Metabolism and Behavior
Faculty of Agriculture, Kyushu University
Fukuoka 812-8511, Japan
Tel/Fax: (81) (92) 642-2953
E-mail: furuse@brs.kyushu-u.ac.jp

In: Proteins Researcher Biographical Sketches ...
Editors: H. Z. Wang and M. Tian

ISBN: 978-1-62100-777-7

Chapter 72

ARGININE REQUIREMENT AND METABOLISM IN MARINE FISH LARVAE- REVIEW OF RECENT FINDINGS

Margarida Saavedra*
Instituto Nacional de Investigação Agrária e das Pescas (INIAP/IPIMAR-CRIPSul), Olhão, Portugal

RESEARCH SUMMARY

Amino acids (AA) are the most important energetic substrate during fish larval development. Estimation of AA requirements is, therefore, crucial to formulate suitable diets for aquaculture species in order to obtain better larval survival, growth and general fish performance. AA larval requirements can be estimated using the AA profile of fish carcass as it is a good indicator of fish larval requirements. The AA profile can later be corrected with the AA bioavailability and a fair estimation of larval AA requirements is achieved. Once the AA requirements are determined they need to be compared to the AA profile of the fish larval diets. Most fish larval stages are still dependent on live feed, such as rotifers and *Artemia,* to survive and grow. Therefore, correction of the AA profiles of the live feed must be done enriching or supplementing the live feed diet. The effect of feed supplementation with AA can be tested at long term through zootechnique trials or at short term using the tube-feeding technique. This review intends to evaluate the current knowledge of arginine requirements and the effect of arginine supplementation on the metabolism of this AA in several marine fish larvae such as white seabream, *Diplodus sargus,* and Senegalese sole, *Solea senegalensis.* Arginine was chosen because it is an indispensable AA involved in metabolic pathways such as protein synthesis, urea production, metabolism of glutamic acid and proline, and synthesis of creatine and polyamines. It is also considered to be in deficiency in the diet of some fish.

* Tel.: +351 289 71 53 46, Fax.: +351 289 71 55 79
E-mail address: margarida.saavedra@gmail.com

In: Proteins Researcher Biographical Sketches ...
Editors: H. Z. Wang and M. Tian
ISBN: 978-1-62100-777-7

Chapter 73

ARGININE-RICH CELL-PENETRATING PEPTIDES IN CELLULAR INTERNALIZATION

Betty Revon Liu[1, 2]***, Huey-Jenn Chiang***[2]
and Han-Jung Lee[1,*]
[1]Department of Natural Resources and Environmental Studies,
National Dong Hwa University, Hualien, Taiwan
[2]Institute of Biotechnology, National Dong Hwa University, Hualien, Taiwan

RESEARCH SUMMARY

Polypeptides composed of arginine-rich domains play a key role in gene regulation, such as *nuclear* localization signal which can penetrate nuclear membrane. Recent studies indicated that arginine-rich cell-penetrating peptides (CPPs) possess the ability to penetrate plasma membrane without receptors. Length of peptides with arginine residues determines the efficiency of cell internalization. Also, *D*- or *L*-form of arginine residues have different effects on internalization efficiency. In our investigations, arginine-rich CPPs could serve as efficient shuttles to deliver proteins, DNAs, RNAs or nanoparticles (such as quantum dots) into animal or plant cells in a covalent (CPT), noncovalent (NPT), or covalent and noncovalent (CNPT) protein transduction manner. The penetrating mechanism of CPPs was still controversial, but more evidences indicated that these peptides enter cells through multiple internalization pathways. No cytotoxicity and high transduction efficiency highlight great advantages of these CPPs in cargoes or drug delivery. Therefore, studies of arginine-rich CPPs could launch a new page in pharmaceutics, therapeutics or cell biology.

[*] Corresponding author: E-mail: hjlee@mail.ndhu.edu.tw

In: Proteins Researcher Biographical Sketches … ISBN: 978-1-62100-777-7
Editors: H. Z. Wang and M. Tian

Chapter 74

EFFECTS OF DEEP SEA WATER ON CHANGES IN FREE AMINO ACIDS AND TOLERANCE TO FUSARIUM ROOT ROT IN MYCORRHIZAL ASPARAGUS PLANTS

Abu Shamim Mohammad Nahiyan[1], Mika Yokoyama[2] and Yoichi Matsubara[2,*]
[1] The United Graduate School of Agricultural Science, Gifu University, Japan
[2] Faculty of Applied Biological Sciences, Gifu University, Japan

RESEARCH SUMMARY

Asparagus (*Asparagus officinalis* L., cv. Welcome) plants inoculated with arbuscular mycorrhizal fungi (AMF) (*Glomus intraradices* and *Gigaspora margarita*). Dry weight of shoots and roots were increased in deep-sea-water added asparagus plants with or without presence of AMF, though higher dry weight was observed in deep-sea-water added AMF-inoculated plots. Thus, plant growth promotion via symbiosis appeared in mycorrhizal asparagus plants. As for disease tolerance, incidence and severity of root rot was lower in AMF plants than in non-AMF plants. On the other hand, several free amino acids , such as asparagine, glutamine, aspartic acid, alanine, GABA, tyrosine, ornithine and lysine were increased in shoots and roots of most of the mycorrhizal asparagus plants. From these findings, it is suggested that plant growth promotion through AMF and tolerance to *Fusarium* root rot occurred in mycorrhizal asparagus plants. In addition, several free amino acid contents are stimulated by the symbiosis, supposing that the changes might be associated with the disease tolerance.

---

* Corresponding author: ymatsu@gifu-u.ac.jp

In: Proteins Researcher Biographical Sketches … ISBN: 978-1-62100-777-7
Editors: H. Z. Wang and M. Tian

Chapter 75

INFLUENCE OF ARGININE-CONTAINING PEPTIDES ON THE HAEMOSTASIS SYSTEM

Maria Golubeva and Marina Grigorjeva
Lab. Blood Coagulation, Biology Dept., Leninsky Gory,
Moscow State University, Moscow, Russia

RESEARCH SUMMARY

It is known that many regulatory peptides take part in haemostatic reactions of the organism, influence on all phases of blood coagulation and fibrinolysis. At the same time formed in process degradation of regulatory peptides fragments have independent biological activity. For example, many short peptides have fibrinolytic, anticoagulant and antitrombotic effects in blood. By stage-by-stage proteolysis it is possible to define structural conditionality of most regulatory peptides effects and to define the minimum fragment, responsible for its biological activity. Any modification of peptide molecule can cause changes of its biological efficiency. Therefore the study of peptides degradation ways, and also structure and effects of formed fragments, has the great interest and significance. In the present study summarizes the results of own researches about the influence on various links of the haemostasis system of same regulatory peptides fragments and different olygopeptides, containing the arginine amino acid in different positions of peptides chain.

In: Proteins Researcher Biographical Sketches ...
Editors: H. Z. Wang and M. Tian

ISBN: 978-1-62100-777-7

Chapter 76

Newly Identified Transcriptional Regulation by Mcm1p at *ARG1* Promoter

Sungpil Yoon*
Research Institute, National Cancer Center, Goyang-si Gyeonggi-do, Republic of Korea

Research Summary

Recent studies with chromation immunoprecipitation assay and mutational analysis in binding sites of the regulators demonstrated unexpected biological mechanisms in the transcriptional regulation of *ARG1*. Two roles of Mcm1p were identified at *ARG1*: a Gcn4p-mediated positive transcriptional role and a negative role involving Arg80p, Arg81p, and Arg82p. Mcm1p contributed to active transcription at the *ARG1* promoter by increasing the binding of the activator Gcn4p and by recruiting the co-activator complex SWI/SNF at *ARG1* under Gcn4p-induced conditions. Mutational analysis of the *ARG1* promoter also revealed a positive role for the Mcm1p binding site in *ARG1* transcription and growth to overcome arginine starvation in the absence of Gcn4p. A transcriptional negative role for Mcm1p was apparent in the recruitment of the whole repressor ArgR/Mcm1p complex, which contributed to dampening the activating function of Gcn4p at *ARG1* in arginine-replete cells. Concerning the mechanism of *ARG1* transcription, the most interesting finding was that *ARG1* transcription could be controlled by different mechanisms with the Mcm1p binding sites through either the presence or absence of complete amino acids under condition of arginine starvation.

* Research Institute, National Cancer Center, 809 Madu 1-dong, Ilsan-gu, Goyang-si Gyeonggi-do, 411-764, Republic of Korea Telephone: (31) 920-2366; Fax: (31) 920-2002, e-mail: yoons@ncc.re.kr

In: Proteins Researcher Biographical Sketches ...
Editors: H. Z. Wang and M. Tian
ISBN: 978-1-62100-777-7

Chapter 77

PROTEIN DISULPHIDE ISOMERASES: DIVERSITY AND ROLES IN PLANTS

Mrinal Bhave**, Huimei Wu and Atul Kamboj
Environment and Biotechnology Centre, Faculty of Life and Social Sciences, Swinburne University of Technology, Hawthorn, Victoria, Australia

RESEARCH SUMMARY

The 'foldase' enzyme protein disulphide isomerase (PDI) interacts with nascent polypeptides in the lumen of the endoplasmic reticulum to catalyze the formation of new disulphide bonds, as well as breakage and alternative isomerization of incorrectly formed bonds, during protein folding and maturation processes in eukaryotic cells. PDI belongs to the thioredoxin (TRX) superfamily that catalyzes cellular redox reactions and has an active site with the conserved motif WCXXC. PDI is a member of the larger PDI-like (PDIL) protein family, its subclasses varying in numbers, positions and sequences of functional domains and active sites and in subcellular locations and functions, many including PDI also playing chaperone roles. The human PDIL family has been studied in great detail, its members exhibiting activities ranging from the archetypical disulphide formation/isomerisation to a multitude of chaperone roles of relevance to cell biology and medicine, for example, in neurodegenerative diseases, apoptosis and cancer. The plant PDIL families are larger and more diverse and have been associated with some unique roles such as storage protein folding; however, information on plant PDILs is much more limited. This work overviews the key properties and functions of human PDILs, summarises the diversity and known functions of plant PDILs, and contemplates the directions for future research on PDILs, particularly in plants.

* Corresponding author: Environment and Biotechnology Centre, Faculty of Life and Social Sciences, Swinburne University of Technology, PO Box 218, Hawthorn, Victoria 3122, Australia, Tel: +61-3-92145759; Fax: +61-3-98190834, E-mail: mbhave@swin.edu.au

In: Proteins Researcher Biographical Sketches ... ISBN: 978-1-62100-777-7
Editors: H. Z. Wang and M. Tian

Chapter 78

SELF-ASSEMBLING PEPTIDES FOR BIOMEDICAL APPLICATIONS: IR AND RAMAN SPECTROSCOPIES FOR THE STUDY OF SECONDARY STRUCTURE

Michele Di Foggia[1], Paola Taddei[1], Armida Torreggiani[2], Monica Dettin[3] and Anna Tinti[1]

[1]Dipartimento di Biochimica – Sez. Chimica e Propedeutica Biochimica, Università di Bologna, Bologna, Italy

[2] Istituto per la Sintesi Organica e la Fotoreattività (I.S.O.F.), Consiglio Nazionale delle Ricerche (C.N.R.), Bologna, Italy

[3] Dipartimento di Processi Chimici dell'Ingegneria, Università di Padova, Padova, Italy

RESEARCH SUMMARY

Self-assembling peptides are a category of peptides which undergo spontaneous assembling into ordered nanostructures. These designed peptides have attracted huge interest in the field of nanotechnology for its potential application in areas such as biomedical nanotechnology, cell culturing, molecular electronics, and more.

In the emerging field of tissue engineering, the development of synthetic materials promoting cell growth has led to the study of regularly alternating polar/non-polar amphiphilic oligopeptides, such as EAK16 $(AEAEAKAK)_2$, also called LEGO peptides, which have been considered particularly promising.

Self-assembling LEGO peptides have a preferential β-sheet structure, are resistant to proteolytic cleavage, and are able to form an insoluble macroscopic membrane under physiological conditions. Their ability to create such stable structures derive from the hydrophobic interactions between the aliphatic groups of non-ionic residues and the complementary ionic bonds between acidic and basic amino acids. This stability can be enhanced by the pH regulation and the presence of monovalent ions.

This chapter will be focused on some approaches useful for elucidating the influence of the sequence modifications and the interactions with a surface on the self-assembly capability of differently synthesised peptides. Eight different oligopeptides (from 16 to 19 residues), derived from EAK16, have been analysed by IR and Raman spectroscopies that are particularly useful for obtaining qualitative and quantitative information on the secondary structure of these peptides. Several modifications in the primary structure of peptides have been considered, namely acidic and/or basic substitutions (Glu → Asp; Lys → Orn), changes in the length of the aliphatic side chain of the spacer residues (Ala → Abu or Ala → Tyr), the addition of RGD (known to promote osteoblast adhesion), or scrambling of the sequence.

As these peptides are widely used as biocompatible coatings of metallic implants, the peptide folding after adsorption on different surfaces, in particular on titanium oxide, will be also discussed.

Finally, it will be shown that not all the oligopeptides examined can self-assemble into a homogeneous multilayer on metallic surfaces; however, most of the peptides take a prevailing β-sheet structure which guarantees the best peptide-surface interaction. In particular, the interactions of polar, ionic and aromatic residues with surfaces will be discussed more in detail.

In: Proteins Researcher Biographical Sketches ...
Editors: H. Z. Wang and M. Tian
ISBN: 978-1-62100-777-7

Chapter 79

STABILITY AND STABILIZATION OF PROTEINS: THE RIBONUCLEASE A EXAMPLE

Ulrich Arnold [1]
Martin-Luther University, Institute of Biochemistry
and Biotechnology, Halle, Germany

RESEARCH SUMMARY

Despite much progress in molecular biology, biophysics, and biochemistry, the mechanisms by which unfolded polypeptide chains gain their native conformation are not yet fully understood. Likewise, the network of forces that stabilize folded proteins and prevent them from unfolding elude a thorough accounting so far. Such an accounting is, however, necessary to engineer protein stability, which is often required for enzymes in technical applications.

The folding process of proteins starts with a hydrophobic collapse or local events at so-called 'nuclei' or 'initiation sites' and proceeds stepwise to the natively folded protein. The occurrence of rate-limiting steps often leads to the population of folding intermediates. In contrast to the successive protein folding process, protein *un*folding is highly cooperative. Due to the principle of microscopic reversibility, unfolding must be a progressive process as well, which is postulated to start at a confined region—'unfolding region', 'critical region' or similar—of the tertiary structure of the native protein molecule.

The rate constants of the folding (k_f) and unfolding reaction (k_U), both of which determine the respective free energy of activation $\Delta G_f^\ddagger$ and $\Delta G_U^\ddagger$, provide information on the contribution of these reactions to the thermodynamic stability (Gibbs free energy $\Delta G = \Delta G_U^\ddagger - \Delta G_f^\ddagger$). The comparison of proteins with their chemically modified or, preferably, genetically engineered variants yields information on the involvement of specific residues in the folding or unfolding reaction.

[1] E-mail address: ulrich.arnold@biochemtech.uni-halle.de

Ribonuclease A (RNase A) has been a model protein for half a century and is one of the most thoroughly studied examples concerning the protein folding problem. Whereas the folding reaction is complex due to proline isomerization reactions, the unfolding reaction can be described by a single-exponential reaction. Modifications by, *e.g.*, genetic engineering have revealed the importance of confined regions for the unfolding and refolding reaction, respectively.

Starting from basic principles of thermodynamics and kinetics, the effect of protein modifications on the stability and tools for protein stabilization are presented with special focus on work on RNase A.

In: Proteins Researcher Biographical Sketches ...
Editors: H. Z. Wang and M. Tian
ISBN: 978-1-62100-777-7

Chapter 80

HETEROLOGOUS PROTEIN FOLDING IN YEAST

Byung-Kwon Choi
GlycoFi, Inc., A Wholly Owned Subsidiary of Merck & Co., Inc.,
Lebanon, New Hampshire, US

RESEARCH SUMMARY

Yeast is an important platform for the production of therapeutic proteins. It has principally been used for the production of relatively simple proteins that lack *N*-linked glycosylation. However, with recent developments in technologies focused on glycoengineering yeast to produce humanized *N*-glycans, its application to the field of human therapeutics becomes greatly potentiated. A constant challenge in the biotechnology industry is the production of recombinant proteins at high levels. Like other protein expression platforms, productivities in yeast depend on the host strain, its process of cultivation and the type of protein being expressed. In addition, it has been demonstrated that yeast lack some molecular chaperones found in higher eukaryotic cells. Protein folding in the endoplasmic reticulum is a key process that can influence the fate of a protein, directing it either to secretion or ER-associated degradation. Some proteins require specific molecular chaperones for their protein maturation and others require general protein folding chaperones. In this review, yeast folding machinery as well as attempts made to improve protein folding will be discussed.

In: Proteins Researcher Biographical Sketches ...
Editors: H. Z. Wang and M. Tian
ISBN: 978-1-62100-777-7

Chapter 81

Modelling of Protein Folding and Prediction of Rate based on Nucleation Mechanism

Oxana V. Galzitskaya*
Institute of Protein Research, Russian Academy of Sciences,
Pushchino, Moscow Region, Russian Federation

Research Summary

The problem of protein self-organization is one of the most important problems of molecular biology nowadays. Despite the recent success in the understanding of general principles of protein folding, details of this process are yet to be elucidated. Moreover, the prediction of protein folding rates has its own practical value due to the fact that aggregation directly depends on the rate of protein folding. The time of folding and transition state ensembles for 67 proteins with known experimental data at the point of thermodynamic equilibrium between unfolded and native state have been calculated using a Monte Carlo method and Dynamic programming one where each residue is considered to be either folded as in the native state or completely disordered. The times of folding for 67 proteins which reach the native state within a limit of 10^8 Monte Carlo steps are in a good correlation with experimentally measured folding time at mid-transition point (the correlation coefficient is -0.82). A lower correlation was obtained if to use Dynamic programming approach (the correlation coefficient is -0.72). The capillarity model allows us to predict the folding rate at the same level of correlation as by Monte-Carlo simulations. The calculated model entropy capacity (conformational entropy per residue divided by the average contact energy per residue) for 67 proteins correlates by about 78% with the experimentally measured folding rate at the mid-transition point. Theoretical consideration of a capillarity model for the process of protein folding demonstrates that the difference in the folding rate for proteins

* Corresponding author: E-mail: ogalzit@vega.protres.ru

sharing more ball-like and less ball-like folds is the result of differences in the conformational entropy due to a larger surface of the boundary between folded and unfolded phases in the transition state for proteins with more ball-like fold.

In: Proteins Researcher Biographical Sketches ... ISBN: 978-1-62100-777-7
Editors: H. Z. Wang and M. Tian

Chapter 82

INTRINSICALLY UNORDERED PROTEINS: STRUCTURAL PROPERTIES, PREDICTION AND RELEVANCE

Susan Costantini[1,2*], Marco Miele[1] and Giovanni Colonna[1**]

[1]Department of Biochemistry and Biophysics and CRISCEB - (Interdepartmental Research Center for Computational and Biotechnological Sciences)
Second University of Naples, Naples, Italy
[2]CROM (Oncology Research Centre of Mercogliano) "Fiorentino Lo Vuolo", Mercogliano, Italy

RESEARCH SUMMARY

The interest in intrinsically unordered proteins (IUPs) has greatly increased, as it has become clear that they are very widespread, especially in eukaryotic organisms. The presence of unordered regions in functional proteins is implicated in important biological roles, such as translation and transcriptional regulation, cell signaling and molecular recognition. A number of studies report that for mammals about 75% of their signalling proteins are predicted to contain long unordered regions (>30 residues), about half of their total proteins are predicted to contain such long unordered regions, and about 25% of their proteins are predicted to be fully unordered. Other studies report examples of unordered proteins implicated in important cellular processes, undergoing transitions to more structured states upon binding to their target ligands, DNA, or other proteins. The accumulation of such evidence led to the suggestion that the classical protein structure-function paradigm needs to be reassessed. In fact, the assumption that a folded three-dimensional structure is necessary for function has

* Corresponding author: via Ammiraglio Bianco, 83013 Mercogliano, Avellino (IT), E-mail: susan.costantini @unina2.it; Tel: +39 0825 1911730; Fax: +39 0825 1911705

** Corresponding author: Department of Biochemistry and Biophysics, Second University of Naples via Costantinopoli 16, 80138 Naples (IT), E-mail: giovanni.colonna@unina2.it; Tel: +39 081 5666574; Fax: +39 081 5665869 been modified. Although the functions of many proteins are directly related to their three-dimensional structures, numerous proteins that lack intrinsic globular structure under physiological conditions are now recognized. As the amino acid sequence contains the information for protein folding, it was reasoned that, also for proteins that do not fold into three-dimensional structures, the residue sequence should specify protein “nonfolding”. Therefore, to test this hypothesis, a large number of methods were developed to predict regions of protein sequences that fail to fold. Their prediction accuracy has been better than expected and this suggests that the information for failure to fold into a regular structure is likely to be inherent within the residue sequence. Because there is an enormous interest in predicting unordered regions of proteins, the analysis of the possibility of predicting protein “unordered” regions was addressed also in the last four editions of CASP experiment.

The main aim of this chapter is to present the state of the art of prediction methods of protein unordered regions developed in the last years and their application to specific protein families. In particular, we will focus our discussion on human Sirt-1 and cytokine membrane receptors because our data suggest that these proteins, involved in chronic inflammatory diseases, have unordered structural segments very important for their function.

In: Proteins Researcher Biographical Sketches ...
Editors: H. Z. Wang and M. Tian
ISBN: 978-1-62100-777-7

Chapter 83

How Do Homodimeric Proteins Fold and Assemble?

Absalom Zamorano-Carrillo*[1], *Jonathan Pablo Carrillo-Vázquez*[1], *Brenda Chimal-Vega*[1], *Oscar Daniel-García*[2], *Roberto Isaac López-Cruz*[1], *Roberto Carlos Maya-Martínez*[1], *Elibeth Mirasol Meléndez*[1] and *Claudia G. Benítez-Cardoza*[1,*]

[1]Laboratorio de Investigación Bioquímica, ENMyH-Instituto Politécnico Nacional. D.F., México

[2]Departamento de Ciencias Naturales. Universidad Autónoma Metropolitana -Cuajimalpa. D. F.; México

Research Summary

Protein folding is an increasingly important field in biomedical research. Tools of physical, chemistry, protein engineering, computer simulation and structural biology have been combined to investigate the folding mechanism of many monomeric proteins.

However, a large percentage of proteins, from all organisms, are oligomeric and most commonly homodimers. In recent years, numerous folding studies have been carried out to understand the folding and assembly pathways of multisubunit proteins and homodimers in particular. This review chapter describes the current knowledge on the thermodynamic and kinetic stability, the folding process and subunit association of homodimeric proteins. The experimental design and interpretation of data from equilibrium and kinetic techniques are revised, which provide practical information for determining the folding pathways. In addition, several folding models are discussed ranging from the simplest two-state model to those that involve one or more intermediate conformations. This information is of interest for

[*] Corresponding autor: Laboratorio de Investigación Bioquímica, Sección de Estudios de Posgrado, ENMyH-IPN. Guillermo Massieu Helguera No. 239, La Escalera Ticoman. D.F. 07320, México Tel.: +52 55 57296300 X 55562; E-mail address: beni1972uk@gmail.com

understanding the folding reaction as well as protein–protein interactions of natural multimeric proteins as well as abnormal complexes implicated in disease.

In: Proteins Researcher Biographical Sketches ... ISBN: 978-1-62100-777-7
Editors: H. Z. Wang and M. Tian

Chapter 84

THE RELATIONSHIP BETWEEN HUMAN MAT1A MUTATIONS AND DISEASE: A FOLDING AND ASSOCIATION PROBLEM?

María A. Pajares[1]* and Claudia Pérez[1,2]
[1]Instituto de Investigaciones Biomédicas "Alberto Sols" (CSIC-UAM), Arturo Duperier, Madrid, Spain
[2]Now at: Instituto de Biotecnología. Universidad Nacional Autónoma de México, Cuernavaca, Morelos, México

RESEARCH SUMMARY

Methionine adenosyltransferases (MATs) are a family of highly conserved oligomers that catalyze the only known reaction for the synthesis of S-adenosylmethionine (AdoMet), the main cellular methyl donor. Their catalytic subunits exhibit a characteristic structure, organized in three domains formed by nonconsecutive stretches of the sequence. The active sites locate at the interface between subunits in the dimer with amino acids of each monomer contributing to catalysis. Changes in activity, oligomerization level and expression have been detected in several hepatic diseases; the knockout mouse for *MAT1A* spontaneously developing hepatocellular carcinoma (HCC). However, none of the patients with persistent hypermethioninemia caused by mutations in this gene exhibits hepatic problems, instead a few cases showing demyelination have been described. This chapter discusses aspects related to the structural features of these enzymes and the impact that the mutations found in the human *MAT1A* gene may have in the final protein structure. The influence of the redox environment in MAT folding and association is also analyzed, in light of the effects that drugs and metals that alter the GSH/GSSG ratio produce in the activity and association level. The recent report of the nuclear localization of the MAT I/III isoenzymes, along with their presence in tissues other than liver opened the option to MAT moonlighting. The possibility exists that disease development is related not only to a decrease in AdoMet production, but

* Corresponding author: Instituto de Investigaciones Biomédicas "Alberto Sols" (CSIC-UAM), Arturo Duperier 4, 28029 Madrid, Spain. (Phone: 34-915854414; FAX: 34-915854401; email: mapajares@iib.uam.es)

also to the role of these particular isoenzymes in different subcellular compartments. Therefore, the influence of *MAT1A* mutations, especially those leading to protein truncations, on folding and subcellular localization is discussed, paying special attention to the Hazelwood's hetero-oligomerization hypothesis to explain the demyelination process in patients with persistent hypermethioninemia.

In: Proteins Researcher Biographical Sketches ... ISBN: 978-1-62100-777-7
Editors: H. Z. Wang and M. Tian

Chapter 85

PARADIGM OF PROTEIN FOLDING IN NEURODEGENERATIVE DISEASES

Pratibha Mehta Luthra*
University of Delhi, Delhi, India

RESEARCH SUMMARY

The most abundant molecules in nature other than water are proteins, which are involved to stimulate or control virtually every chemical process. Majority of proteins must be converted into tightly folded compact structures in order to function and their ability to fold to a unique structure generate enormous selectivity and diversity in their functions. Conformational changes in proteins result in misfolding, aggregation, and intra- or extraneuronal accumulation of amyloid fibrils in many neurodegenerative disorders. Proteins appear to fold by diverse pathways, the rates of folding are largely determined by the topology of the native structure. In general, secondary structure is inherently unstable and its stability is enhanced by tertiary interactions. The most conspicuous feature of many neurodegenerative disorders, including Alzheimer's, Parkinson's, and Huntington's disease, is the occurrence of protein aggregates in ordered fibrillar structures known as amyloid found inside and outside of brain cells. The appearance of aggregates in diseased brains implies an underlying incapacity in the cellular machinery of molecular chaperones that normally functions to prevent the accumulation of misfolded proteins. Molecular chaperones provide a first line of defense against misfolded, aggregation-prone proteins, and are among the most potent suppressors of neurodegeneration known for animal models of human disease. A detailed understanding of the molecular basis of protection by chaperones against neurodegeneration might lead to the development of therapies for neurodegenerative disorders that are associated with protein misfolding and aggregation.

In this review, the advances on the protein folding in neurodegenrative disorder and unfolding the role of molecular chaperones in neuroprotection has been discussed.

---

* Division of Medicinal Chemistry, Dr. B.R. Ambedkar Centre for Biomedical Research, University of Delhi, Delhi 110007, India, Phone: 91-11-27666272, Fax : 91-11-27666248, Email: pmluthra@acbr.du.ac.in, pmlsci@yahoo.com

In: Proteins Researcher Biographical Sketches ...
Editors: H. Z. Wang and M. Tian
ISBN: 978-1-62100-777-7

Chapter 86

DIFFERENTIAL SCANNING CALORIMETRY: THERMODYNAMIC ANALYSIS OF THE UNFOLDING TRANSITIONS OF PROTEINS, DOMAINS AND PEPTIDIC FRAGMENTS BY USING EQUILIBRIUM MODELS

Jose C. Martinez, Eva S. Cobos, Irene Luque and Javier Ruiz-Sanz
Department of Physical Chemistry and Institute of Biotechnology, Faculty of Sciences, University of Granada, Granada, Spain

RESEARCH SUMMARY

The great advance achieved during the last decade in the structural, dynamic, energetic and functional knowledge of proteins has been mainly due to the parallel development of instruments and methodologies to interpret experimental data. The special attention may have brought the progress into the *Protein Engineering and Biotechnology* fields. Thus, the current availability of macromolecules *"a la carte"* has allowed researchers to deeply understand the second half of the genetic code, this is, how a certain aminoacidic sequence derives, under defined conditions, into the native state of such protein. The answer to this open question will require the establishment of basic *folding rules* to rationally design *de novo* protein sequences with well defined functionalities, to be used for medical and pharmacological purposes, for the development of bioreactors, biosensors, etc. In other words, these studies may lead to a truly industrial and scientific revolution.

The folding process of proteins is described by a series of more or less complex mechanisms, mainly depending on the nature and, in general, on the size and structural complexity of each protein. In this way, the folding studies carried out with low-size proteins have granted a much more simplistic and informative analysis of calorimetric and spectroscopic information, which has permitted a better understanding of the folding essentials and a tangible advance towards rational design. From the diverse methodologies that can be applied in the research of protein folding processes, differential scanning calorimetry (DSC) presents a series of advantages and possibilities, such technique being considered as a very powerful tool. Precluding the structural interpretation, the direct determination of folding thermodynamic parameters becomes necessary to describe the energetic aspects of the folding and, thus, to define and to rationalize protein stability. Nevertheless, although calorimetry has been widely used as an experimental resource, it has not always been correctly interpreted, mainly because of the difficulty found in extracting thermodynamic information from experimental data. Thus, the rigorous analysis of DSC traces should be done under the assumption of theoretical models, able to describe the most significant stages present during the unfolding process, and which application would give rise to valuable thermodynamic information for each of such stages.

In this chapter we intend to describe in detail the most usual DSC equilibrium and kinetic models that explain the folding behavior of small proteins, domains and peptidic fragments, paying special attention to the concrete experimental and computational aspects that should be taken into account to obtain as much thermodynamic information as possible. We will also explain the best strategies to properly discern between the different folding models and to improve the accuracy of the respective thermodynamic parameters.

In: Proteins Researcher Biographical Sketches ... ISBN: 978-1-62100-777-7
Editors: H. Z. Wang and M. Tian

Chapter 87

STUDY OF FOLDING/UNFOLDING KINETICS OF LATTICE PROTEINS BY APPLYING A SIMPLE STATISTICAL MECHANICAL MODEL FOR PROTEIN FOLDING

Hiroshi Wako[1] and Haruo Abe[2]
[1]School of Social Sciences, Waseda University, Tokyo, Japan
[2]Department of Digital Engineering, Nishinippon Institute of Technology, Fukuoka, Japan

RESEARCH SUMMARY

The folding/unfolding kinetics of a three-dimensional lattice protein was studied using a simple statistical mechanical model for protein folding that we had developed earlier. The model considers the specificity of an amino acid sequence and the native structure of a given protein. We calculated the characteristic relaxation rate on the free energy surface starting from a completely unfolded structure (or native structure) that is assumed to associate with a folding rate (or an unfolding rate). The chevron plot of these rates as a function of the inverse temperature was obtained for four lattice proteins, *a1*, *a2*, *b1*, and *b2*, in order to investigate the dependency of the folding and unfolding rates on their native structures and amino acid sequences. Proteins *a1* and *a2* fold to the same native structure, but their amino acid sequences differ. The same is true for proteins *b1* and *b2*, but their native structure is different from that of *a1* and *a2*. To elucidate the roles of individual amino acid residues in protein folding/unfolding kinetics, we calculated the kinetic properties for all possible single amino acid substitutions of these proteins and examined their responses. The results are discussed with respect to the roles of short- and long-range interactions and formation of a folding nucleus in the kinetics of protein folding/unfolding.

In: Proteins Researcher Biographical Sketches ...
Editors: H. Z. Wang and M. Tian
ISBN: 978-1-62100-777-7

Chapter 88

RIBOSOME ASSISTED PROTEIN FOLDING: SOME OF ITS BIOLOGICAL IMPLICATIONS

Dibyendu Samanta[2,3]***, Anindita Das***[2]***, Debasis Das***[2]***,***
Arpita Bhattacharya[2]***, Arunima Basu***[2]***, Jaydip Ghosh***[2]
and Chanchal DasGupta[1,2*]
[1]Department of Biological Sciences; IISER-Kolkata, NITTTR Campus, Salt Lake, Kolkata, India
[2]Department of Biophysics, Molecular Biology and Genetics; University College of Science; Kolkata, India
[3]Department of Microbiology and Immunology, Albert Einstein College of Medicine, US

RESEARCH SUMMARY

One of the central characteristics of a living system is the ability to self-assemble its component molecular structures with precision and fidelity. The folding of proteins into their compact three-dimensional structures is the most fundamental and universal example of biological self-assembly. Understanding this complex process will therefore make available a unique insight into the way in which evolutionary selection has influenced the properties of a molecular system for functional advantage. The final goal of folding studies is to predict structure from sequence, allowing the design of new functional proteins and prevention of abnormal disease-associated protein conformations.

* Corresponding author: chanchaldg2000@yahoo.com or dibyendu.samanta@einstein.yu.edu

In: Proteins Researcher Biographical Sketches ... ISBN: 978-1-62100-777-7
Editors: H. Z. Wang and M. Tian

Chapter 89

BACTERIAL CYCLOPHILINS

***Angel Manteca*[1, 2] *and Jesus Sanchez*[1]**
[1]Area de Microbiologia, Departamento de Biologia Funcional and IUBA, Facultad de Medicina, Universidad de Oviedo, Oviedo, Spain
[2]Protein Research Group, Department of Biochemistry and Molecular Biology, University of Southern Denmark, Campusvej, DK-Odense M, Denmark

RESEARCH SUMMARY

Peptidyl-prolyl cis/trans isomerases (PPIases; EC 5.2.1.8) are folding helper enzymes which accelerate the formation of the cis/trans equilibrium of peptidyl-prolyl bonds, a rate-limiting process during protein folding events. PPIases are ubiquitous proteins present in almost all living organisms. They can be classified into three structurally and biochemically divergent subfamilies: cyclophilins, FKBP (FK506 binding proteins), and parvulins. No sequence homology exists between the three families, and they have characteristic substrate specificities. In this chapter, we review the current knowledge about bacterial cyclophilins: biochemical characteristics, diversity, and evolutive relationships.

In: Proteins Researcher Biographical Sketches ...
Editors: H. Z. Wang and M. Tian
ISBN: 978-1-62100-777-7

Chapter 90

COMPLETE DESCRIPTION OF PROTEIN FOLDING SHAPES FOR STRUCTURAL COMPARISON

Jiaan Yang*
Sundia Meditech Company, Ltd. Shanghai, China

RESEARCH SUMMARY

The effort to develop better description of protein three-dimensional folding structures has dominated biochemistry and drug discovery research for more than 70 years since Pauling first defined the helical configurations as secondary structure for protein in 1940. The challenge is how to acquire a complete description of protein folding shapes from N-terminal to C-terminal, including regular secondary structure as well as irregular tertiary structure. Here, a novel description method is introduced, which a set of 27 vectors is rigorously derived mathematically from an enclosed space. Each vector represents a three-dimensional folding shape of five successive C_α atoms, and the protein conformation can be completely described along protein backbone. These vectors are expressed by 27 alphabetic symbols, which are called as protein folding shape code (PFSC). Consequently, with PFSC, the folding conformation of any protein with given three-dimensional structure is able to be converted into a simple one-dimensional alphabetic string without gap. Furthermore, to take the advantage of one-dimensional description of folding shapes, the protein conformational structures are able to be compared with Needleman-Wunsch alignment algorithm. The global similarity of protein 3D structures is able to be assessed by a value of protein folding structure alignment score (PFSA-S) as a quantitative measurement, and the similarity and dissimilarity of local structures is able to be examined by alignment table. The results show that this approach has the capability not only to distinguish protein conformers with relatively high similarity, but also to compare proteins with diverse degrees in structural homology. Therefore, this approach provides a consistent procedure, and it produces a unique score for assessment of similarity in protein structure comparison. The significant is that the complete

* Corresponding author: Email: jyang@sundia.com; Phone: 86-21-51098642

description of protein folding shapes provides a simple and effective means to screen protein database, compare protein structures, search protein fragment and probe drug binding site, study protein mutation and protein misfolding and so on.

In: Proteins Researcher Biographical Sketches … ISBN: 978-1-62100-777-7
Editors: H. Z. Wang and M. Tian

Chapter 91

FOLDING AND UNFOLDING OF HYPERTHERMOPHILIC PROTEINS; MOLECULAR BASIS OF ADAPTATION TO HOT ENVIRONMENT

Atsushi Mukaiyama and Kazufumi Takano*

[1]Division of Biological Science, Graduate School of Science,
Nagoya University, Furo-cho, Chikusa-ku, Nagoya
[2]Department of Material and Life Science, Osaka University,
Yamadaoka, Suita, Osaka, Japan

RESEARCH SUMMARY

Many organisms grow in extreme environments on Earth. Microorganisms whose optimal growth temperature is above 80°C are called hyperthermophiles. Proteins from hyperthermophiles usually exhibit higher stability than those from organisms that grow at lower temperatures. In this paper, we describe our studies of the stability and folding of ribonuclease HII from the hyperthermophile *Thermococcus kodakaraensis* and introduce other studies of hyperthermophilic proteins. These studies focus on hyperthermophilic proteins with reversible unfolding and examine their equilibrium and kinetic aspects. The results indicate that kinetic stability (slow unfolding) is a general strategy by which hyperthermophilic proteins adapt to higher temperatures. Mutational analysis indicates that hydrophobic interaction is a factor in the molecular mechanism of the slow unfolding of hyperthermophilic proteins. Next, we report on studies of subtilisin-like serine protease from *Thermococcus kodakaraensis*. The results demonstrate that this protein is extremely stable and that the maturation (folding) process of this protein differs from those of mesophilic homologues in terms of the role of Ca^{2+} and propeptide. This protein is thought to be stabilized due to a high kinetic barrier between the unfolded and folded states. To reduce this barrier, the folding may be fine-tuned by utilizing Ca^{2+} and propeptide. We discuss protein folding from the viewpoint of adaptation to hot environments.

* Corresponding author: Email: J ktakano@mls.eng.osaka-u.ac.jp

In: Proteins Researcher Biographical Sketches ... ISBN: 978-1-62100-777-7
Editors: H. Z. Wang and M. Tian

Chapter 92

DECODING AMINO ACID SEQUENCES OF PROTEINS USING INTER-RESIDUE AVERAGE DISTANCE STATISTICS TO EXTRACT INFORMATION ON PROTEIN FOLDING MECHANISMS

Takeshi Kikuchi*
Department of Bioinformatics, College of Biosciences,
Ritsumeikan University, Japan

RESEARCH SUMMARY

One of the ultimate goals of bioinformatics is to discover the principles governing protein folding. However, information on folding mechanisms of proteins cannot be easily decoded from amino acid sequences using standard bioinformatics techniques such as multiple sequence alignment and so on. We still do not sufficiently understand how the determinants of protein folding are encoded in amino acid sequences. In this review, we summarize how information about protein folding mechanisms can be extracted from amino acid sequences using inter-residue average distance statistics. Our kind of predicted contact map (Average Distance Map, ADM) pinpoints regions of possible folding nuclei for proteins. The amino acid sequence of a given protein suffice for construction of the ADM of a protein. An inter-residue effective potential is defined from the average distance statistics and is utilized to predict folding nucleation sites of a protein sequence. We review examples of the applications of these methods to the analyses of the protein folding of members of several protein families. The general possibility of predicting protein folding mechanisms from only sequences is also discussed.

* Corresponding author: e-mail: tkikuchi@sk.ritsumei.ac.jp 1-1-1 Nojihigashi, Kusatsu, Shiga 575-8577, JAPAN

In: Proteins Researcher Biographical Sketches ...
Editors: H. Z. Wang and M. Tian

ISBN: 978-1-62100-777-7

Chapter 93

REDOX-DEPENDENT CHAPERONING, FOLLOWING PDI FOOTSTEPS

Olivier Serve[1,2], Yukiko Kamiya[1,2] and Koichi Kato[1,2]
[1]Okazaki Institute for Integrative Bioscience and Institute for Molecular Science,
National Institutes of Natural Sciences, Higashiyama,
Myodaiji, Okazaki, Aichi, Japan
[2]Graduate School of Pharmaceutical Sciences, Nagoya City University,
Tanabe-dori, Mizuho-ku, Nagoya, Japan

RESEARCH SUMMARY

Protein disulfide isomerase (PDI) is a protein-folding assistant and a molecular chaperone found in the endoplasmic reticulum. This enzyme catalyzes the formation and the isomerization of disulfide bonds. The binding and release of substrate from molecular chaperones are usually triggered by cycles of ATP hydrolysis. It has been proposed that the substrate binding / release cycle of PDI could be correlated with the redox states of its active sites. The idea that PDI might act as redox-dependent molecular chaperone opened a new path in the field of chaperoning.

However, this redox dependence has not yet been fully established and remains subjected to controversy. We wish to address this issue here on the basis of the current knowledge on PDI including our recent findings on conformational change of thermophilic fungal PDI. Toward this end, based on inspection of the characteristics of the modular structure of PDI, we present in this chapter the differences between the domains that compose PDI in regards to their functions and the consequences of the conformational alteration they experience upon redox condition changes. Finally, the mechanistic model we propose for the action of PDI supports the hypothesis of redox-dependent chaperoning.

In: Proteins Researcher Biographical Sketches ...
Editors: H. Z. Wang and M. Tian ISBN: 978-1-62100-777-7

Chapter 94

ON THE MYOGLOBIN FOLDING IN ORGANIC SOLVENTS AND COSOLVENTS

Katia C. S. Figueiredo**, Helen C. Ferraz, Cristiano P. Borges and Tito L. M. Alves
Programa de Engenharia Quimica – COPPE – Universidade Federal do Rio de Janeiro
Centro de Tecnologia, Ilha do Fundao – Rio de Janeiro, RJ – Brazil

RESEARCH SUMMARY

Our goal in this work was the evaluation of the structural stability of horse heart myoglobin in organic solvents and cosolvents. The selection of an organic solvent or solvent mixture able to maintain the structural stability of myoglobin in organic phase can make viable the use of such media to replace the aqueous one. Changes in the structure of myoglobin were monitored in spectroscopic tests, UV-visible and circular dichroism, CD. The former was used to infer the changes in heme group as a result of the addition of organic solvent whereas the latter was performed to assess the secondary structure of the protein. The increase in the solubility of myoglobin in nonpolar solvents was investigated by means of hydrophobic ion pairing. Sodium dodecyl sulfate, SDS, was used to replace small counter ions on the protein surface and change the miscibility in nonpolar solvents. The results of myoglobin in alcohols indicated that the increase in hydrocarbon content and the decrease in alcohol branching increased their denaturing ability. The most interesting result was the dissolution of myoglobin in pure methanol, which corroborates the molecular dynamic simulation of proteins in this organic solvent. In general, the UV-visible spectra revealed that polar organic solvents have better performance in the dissolution of myoglobin than the polar ones. Protic solvents, like alcohols and glycols, exhibited low denaturing ability compared to the aprotic ones. This behavior can be reasoned in terms of the interaction between the solvent and the protein. Protic solvents probably establish hydrogen bonds with the protein polar groups, similar to water. However, solvents with high polarity can damage the native folding of proteins, because of the removal of water shell. CD spectroscopy was used as a

* Corresponding author: Email: katia@peq.coppe.ufrj.br. Tel: +55 21 2562 8343 / Fax: +55 21 2562 8300

complementary test for the systems with unaltered UV-visible absorbance profile to infer the secondary structure of the protein. The CD results showed the potential of glycols, namely glycerol and ethylene glycol, as cosolvents of myoglobin, up to 50% in volume. Regarding the change in myoglobin superficial moiety, the absorbance profile was maintained for the molar ratio SDS/myoglobin of 5, showing that the protein was effectively extracted to the organic phase (hexane) as the result of SDS pairing on its surface. Higher content of SDS led to turbidity. This result showed the potential of the technique to increase the solubility of myoglobin in nonpolar solvents.

However, this redox dependence has not yet been fully established and remains subjected to controversy. We wish to address this issue here on the basis of the current knowledge on PDI including our recent findings on conformational change of thermophilic fungal PDI. Toward this end, based on inspection of the characteristics of the modular structure of PDI, we present in this chapter the differences between the domains that compose PDI in regards to their functions and the consequences of the conformational alteration they experience upon redox condition changes. Finally, the mechanistic model we propose for the action of PDI supports the hypothesis of redox-dependent chaperoning.

In: Proteins Researcher Biographical Sketches ... ISBN: 978-1-62100-777-7
Editors: H. Z. Wang and M. Tian

Chapter 95

FUNCTIONAL SIGNIFICANCE OF INTRINSICALLY DISORDERED CONFORMATIONS IN ACTIVATION DOMAINS OF THE TRANSCRIPTION FACTORS

R. Kumar*
Department of Basic Sciences, The Commonwealth Medical College, Scranton, PA, US

RESEARCH SUMMARY

In spite of well established fact that gene regulation by the transcription factors (TFs) involves the regulated assembly of multi-protein complexes on enhancers and promoters, scientific community is still actively searching for answers to questions pertaining to the mechanisms that govern this complex process. In this context the structural determinant of activation domains (ADs) of TFs, which provide a basic platform for the assembly of other coregulatory proteins (critical in gene regulation), is of immense importance. It has become quite clear that many TFs possess AD(s), which exist in an intrinsically disordered (ID) conformation. In recent years, it has been shown that significant biological functions are associated with ID states of proteins. Among others, the flexibility of intrinsic disorder in ADs facilitates post-translational modifications, adoption of different conformations with various binding partner proteins, which in turn may lead to its diverse role in gene regulation by the TFs. Thus, proper understanding of ID AD regions in the action of signaling molecules may lead to a new approach for drug discovery, targeted at the assembly of multi-protein complexes involving TFs. This article reviews the available understanding of disorder-order transition in the gene regulation by the TFs, and provides future perspectives for the utilization of this knowledge for novel approaches to drug discovery.

* Corresponding author: Email: rkumar@tcmedc.org, Phone: 570-504-9675 Fax: 570-504-9660

In: Proteins Researcher Biographical Sketches … ISBN: 978-1-62100-777-7
Editors: H. Z. Wang and M. Tian

Chapter 96

COARSE GRAINED PROTEIN MODELING

Carlo Guardiani[1,*] and Fabio Cecconi[2,†]
[1] Centro Interdipartimentale per lo Studio delle Dinamiche Complesse (CSDC)
Sezione INFN di Firenze, Italy
[2] Istituto dei Sistemi Complessi ISC-CNR, UoS "Sapienza", Rome

RESEARCH SUMMARY

Coarse-grained models are experiencing a renewed interest as they grant access to biologically relevant length and time-scales, they allow to single out molecular driving forces from a plethora of biochemical details. Topological models, discussed in the first part of the chapter, are based on the minimal frustration principle that justifies the stability and fast accessibility of the native state. Despite the several underlying approximations, this rich class of models also including Ising-like and Elastic Network Models, has been successfully applied in several contexts, from biomedical problems to the study of mechanical unfolding and translocation. Conversely, the necessity to elucidate the role of the amino-acid sequence on protein folding, in agreement with Anfinsen's postulates, inspires sequence-based models that are reviewed in the second part of the survey. As discussed, each model presents advantages and limitations and the choice of a model is dependent on the problem that must be addressed.

* E-mail address: guardiani@fi.infn.it
† E-mail address: fabio.cecconi@roma1.infn.it

In: Proteins Researcher Biographical Sketches ... ISBN: 978-1-62100-777-7
Editors: H. Z. Wang and M. Tian

Chapter 97

PROTEIN FOLDING: A PERSPECTIVE FROM STATISTICAL PHYSICS

Jinzhi Lei[1,*] and Kerson Huang[2,3]
[1]Zhou Pei-Yuan Center for Applied Mathematics, Tsinghua University, Beijing, China
[2]Physics Department, Massachusetts, Institute of Technology, Cambridge, MA, US
[3]Institute of Advanced Studies, Nanyang Technological University, Singapore

RESEARCH SUMMARY

In this chapter, we introduce an approach to the protein folding problem from the point of view of statistical physics. Protein folding is a stochastic process by which a polypeptide folds into its characteristic and functional 3D structure from random coil. The process involves an intricate interplay between global geometry and local structure, and each protein seems to present special problems. The first part of this chapter contains a concise discussion on kinetics versus thermodynamics in protein folding, and introduce the statistical physics basis of protein folding. In the second part, we introduce CSAW (conditioned self-avoiding walk), a model of protein folding that combines the features of self-avoiding walk (SAW) and the Monte Carlo method. In this model, the unfolded protein chain is treated as a random coil described by SAW. Folding is induced by hydrophobic forces and other interactions, such as hydrogen bonding, which can be taken into account by imposing conditions on SAW. Conceptually, the mathematical basis is a generalized Langevin equation. Despite the simplicity, the model provides clues to study the universal aspects while we overlook details and concentrate only on a few general properties. To illustrate the flexibility and capabilities of the model, we consider several examples, including helix formation, elastic properties, and the transition in the folding of myoglobin. From the CSAW simulation and physical arguments, we find a universal elastic energy for proteins, which depends only on the radius of gyration Rg and the residue number N. The elastic energy gives rise to scaling laws $Rg \sim N^{\nu}$ in different regions with exponents $\nu = 3/5, 3/7, 2/5$.

* E-mail address: jzlei@mail.tsinghua.edu.cn

In: Proteins Researcher Biographical Sketches ... ISBN: 978-1-62100-777-7
Editors: H. Z. Wang and M. Tian

Chapter 98

PROTEOMICS IN CELIAC DISEASE

V. De Re, MP. Simula, L. Caggiari, A. Pavan , V. Canzonieri and R. Cannizzaro
Farmacologia Sperimentale e Clinica DOMERT,
Centro di Riferimento Oncologico, IRCCS, Aviano, Pordenone, Italy

RESEARCH SUMMARY

Proteomic technologies are used with increasing frequency in the scientific community. In this review we would like to highlight their use in celiac disease. The available techniques that include two-dimensional gel electrophoresis, mass spectrometry, antibody and tissue arrays, have been used to identify proteins or protein expression changes specific of gut tissue from patients with celiac disease. A number of studies have employed proteomic methodologies to look for diagnostic biomarkers in body fluids or to examine protein expression changes and posttranslational modifications during signaling. The fast technological development of technologies, along with the combination of classic techniques with proteomics, will lead to new discoveries which will consent a better understanding of the pathogenesis of celiac disease and its complications (i.e. refractory CD and cancer), and to possibily indicate targets for an early diagnosis of CD complications and for specific terapeutic approaches.

In: Proteins Researcher Biographical Sketches …
Editors: H. Z. Wang and M. Tian
ISBN: 978-1-62100-777-7

Chapter 99

THE PUZZLE OF PROTEIN LOCATION IN PLANT PROTEOMICS

Elisabeth Jamet and Rafael Pont-Lezica
Surfaces Cellulaires et Signalisation chez les Végétaux,
Université de Toulouse, Pôle de Biotechnologie Végétale,
Castanet-Tolosan, France

RESEARCH SUMMARY

Organelle proteomics allows the characterization of complex proteomes to understand the protein networks which regulate growth and development, as well as adaptation and evolution. Purification of organelles is of paramount importance and diverse protocols are published. Some organelles such as chloroplasts, mitochondria, and the nucleus are surrounded by membranes which facilitate their purification. Others have membranes easily disrupted (vacuoles and peroxisomes), or are complex systems for protein trafficking (endoplasmic reticulum, Golgi, and secretory vesicles). The cell walls present different difficulties since they have no physical limits allowing purification. The purity of the targeted cell compartment is usually evaluated by biochemical and/or immunological methods. Nevertheless, in any sub-cellular proteomic analysis, proteins from a different compartment can be detected and the difficulty is to decide whether it is a contamination, or the unexpected location is real and has a functional significance. Software to predict sub-cellular location of proteins is available. However, since not all the targeting signals are known at present, carefulness in the use of such tools is recommended. Different tactics to solve this puzzle are discussed in this commentary.

In: Proteins Researcher Biographical Sketches ...
Editors: H. Z. Wang and M. Tian

ISBN: 978-1-62100-777-7

Chapter 100

WHAT FUTURE FOR "GEL-BASED PROTEOMIC" APPROACHES?

François Chevalier*
Proteomic Laboratory, iRCM, CEA, Fontenay aux Roses, France

RESEARCH SUMMARY

The simultaneous analysis of all proteins expressed by a cell, tissue or organism in a specific physiological condition is the main goal of proteomic studies. Gel-based proteomic is the most popular and versatile method of global protein separation and quantification. This is a mature approach to screen the protein expression at the large scale. Based on two independent biochemical characteristics of proteins, two-dimensional electrophoresis combines isoelectric focusing, which separates proteins according to their isoelectric point, and SDS-PAGE, which separates them further according to their molecular mass. The next typical steps of the flow of gel-based proteomics are spots visualization and evaluation, expression analysis and finally protein identification by mass spectrometry. At present, two-dimensional electrophoresis allows simultaneously to detect and quantify up to thousand protein spots in the same gel in a wide range of biological systems for the study of differentially expressed proteins. However, gel-based proteomic has a number of inherent drawbacks. In this review article, the benefits, difficulties, limits and perspectives of gel-based proteomic approaches are discussed.

* Tel: +33 (0)146 548 326, Fax: +33 (0)146 549 138, Email: francois.chevalier@cea.fr

In: Proteins Researcher Biographical Sketches ...
Editors: H. Z. Wang and M. Tian
ISBN: 978-1-62100-777-7

Chapter 101

ALGORITHMS FOR THE QUANTIFICATION OF PROTEINS FROM HIGH-THROUGHPUT LIQUID CHROMATOGRAPHY-MASS SPECTROMETRY (LC-MS) DATA

Ole Schulz-Trieglaff
Berlin, Germany

RESEARCH SUMMARY

In this commentary, we review algorithms for the analysis of liquid chromatography-mass spectrometry (LC-MS) data. Mass spectrometry is a technology that can be used to determine the identities and abundances of the compounds in complex samples. In combination with liquid chro-matography, it has become a popular method in the field of proteomics, the large-scale study of proteins and peptides in living systems. The data sets obtained from an LC-MS experiment are large and highly complex. The outcome of such an experiment is called an LC-MS map and is a collection of mass spectra. They contain, among the signals of interest, a high amount of noise and other disturbances. That is why algorithms for the low-level processing of LC-MS data are becoming increasingly important. In this commentary, we revied the state-of-the-art of quantification algorithms, their capabilities and also limitations and outline avenues for future research.

In: Proteins Researcher Biographical Sketches ...
Editors: H. Z. Wang and M. Tian
ISBN: 978-1-62100-777-7

Chapter 102

METHOD FOR PREDICTION OF PROTEIN-PROTEIN INTERACTIONS IN YEAST USING GENOMICS / PROTEOMICS INFORMATION AND FEATURE SELECTION

J. M. Urquiza, I. Rojas, H. Pomares and L. J. Herrera*
Department of Computer Architecture and Computer Technology,
University of Granada, Granada, Spain

RESEARCH SUMMARY

Nowadays, one of the most important goals of Proteomics is the prediction of protein-protein interactions (PPIs), whose knowledge is vital for all biological processes. In the present paper we propose an approach to the prediction of protein-protein interactions in yeast based on the well-known paradigm of Support Vector Machines (SVM) for classification and feature selection methods using Genomics/Proteomics information from the main databases. In order to obtain higher values of specificity and sensitivity for our predictions, we took a high reliable set of positive and negative examples for the construction of the SVM model. We then extracted a set of proteomic/genomic features from these examples and also introduced a similarity measure in the calculation of the features, that allows us to improve the prediction capability of our model. In the analysis of the results, we also applied our approach to in vitro datasets, obtaining high accuracy classifications. Our final SVM classifiers obtain a low error rate in the prediction for each pair of proteins of several datasets for both in vitro and in silico methodologies.

* E-mail address: (jurquiza, irojas, hpomares, jherrera)@atc.ugr.es

In: Proteins Researcher Biographical Sketches ...
Editors: H. Z. Wang and M. Tian
ISBN: 978-1-62100-777-7

Chapter 103

LABEL-FREE LIQUID CHROMATOGRAPHY-BASED QUANTITATIVE PROTEOMICS: CHALLENGES AND RECENT DEVELOPMENTS

A. Matros*[*1], *S. Kaspar*[1], *S. Tenzer*[2], *M. Kipping*[3], *U. Seiffert*[4] and *H.-P. Mock*[1]

[1]Leibniz Institute of Plant Genetics and Crop Plant Research, Gatersleben, Germany
[2]Institute of Immunology, University of Mainz, Mainz, Germany
[3]Waters GmbH, Eschborn, Germany
[4]Fraunhofer-Institute IFF Magdeburg, Biosystems Engineering, Magdeburg, Germany

RESEARCH SUMMARY

Recent innovations in liquid chromatography-mass spectrometry (LC-MS) based methods have facilitated comparative and functional proteomic analyses of large numbers of proteins derived from complex samples without any need for protein or peptide labelling. Here we discuss the features of label-free LC-based proteomics techniques. We first summarize recent methods used for quantitative protein analyses by MS techniques. The major challenges faced by label-free LC-MS based approaches are discussed; these include sample preparation, peptide separation, data mining and quantification. Absolute quantification, kinetic approaches and database search algorithms are also addressed. We focus on the Expression^{E} System^{TM} (Waters, Manchester, UK), a relatively new platform allowing label-free quantification of peptides for which mass and retention time have been accurately measured. Enhancing the power of this method will require developments in both separation technology and bioinformatics/statistical analysis.

* Corresponding Author: Andrea Matros, Leibniz Institute of Plant Genetics and Crop Plant Research, Dept. Molecular and Cell Physiology, Applied Biochemistry Group, Corrensstrasse 3, D-06466, Gatersleben, Germany, Phone: 49 39482 5-445, Fax:+49 39482 5-524

In: Proteins Researcher Biographical Sketches …
Editors: H. Z. Wang and M. Tian
ISBN: 978-1-62100-777-7

Chapter 104

INSIGHTS FROM PROTEOMICS INTO MILD COGNITIVE IMPAIRMENT, LIKELY THE EARLIEST STAGE OF ALZHEIMER'S DISEASE

Renã A. Sowell and D. Allan Butterfield
University of Kentucky, Lexington, KY, US

RESEARCH SUMMARY

Mild cognitive impairment (MCI) is arguably the earliest form of Alzheimer's disease (AD). Better understanding of brain changes in MCI may lead to the identification of therapeutic targets to slow the progression of AD. Oxidative stress has been implicated as a mechanism associated with the pathogenesis of both MCI and AD. In particular, among other markers, there is evidence for an increase in the levels of protein oxidation and lipid peroxidation in the brains of subjects with MCI. Several proteins are oxidatively modified in MCI brain, and as a result individual protein dysfunction may be directly linked to these modifications (e.g., carbonylation, nitration, modification by HNE) and may be involved in MCI pathogenesis. Additionally, Concanavalin-A-mediated separation of brain proteins has recently led to the identification of key proteins in MCI and AD using proteomics methods. This chapter will summarize important findings from proteomics studies of MCI, which have provided insights into this cognitive disorder and have led to further understanding of potential mechanisms involved in the progression of AD.

In: Proteins Researcher Biographical Sketches ... ISBN: 978-1-62100-777-7
Editors: H. Z. Wang and M. Tian

Chapter 105

MULTIDIMENSIONAL CHROMATOGRAPHY: AN ESSENTIAL TOOL FOR PROTEOMICS

Chiara Cavaliere, Eleonora Corradini, Patrizia Foglia, Piero Giansanti, Roberto Samperi and Aldo Laganà *
Department of Chemistry, SAPIENZA Università di Roma, Rome, Italy

RESEARCH SUMMARY

The general strategy in proteomic research consists in sample preparation, protein or peptide separation, their identification, and data interpretation. A critical step is certainly protein or peptide separation. Since increasingly complicated biological structures are studied by mass spectrometry (MS), the need for more powerful and highly resolving separation methods is growing. Consequently, multidimensional separation techniques in combination with MS have emerged as a powerful tool for the large-scale proteomic analysis. Until recently, two dimensional gel electrophoresis (2-DE) was the technique most often used for protein separation. The limitations of 2-DE in detecting low abundance, very small or large proteins, basic and membrane/hydrophobic ones, as well as difficulties with process automation, have forced researchers to look for other methods of protein separation, such as multidimensional liquid chromatography coupled to MS (MDLC-MS) or tandem MS (MDLC-MS/MS). MDLC combines two or more forms of LC to increase the peak capacity, and thus the resolving power of separation, to better fractionate peptides prior to entering the mass spectrometer. In this chapter, we analyze status and recent developments of the MDLC experiments in their fundamental components. It describes a variety of separation modes that have been employed to achieve protein-level or peptide-level separation, including size exclusion chromatography, ion exchange chromatography, and reversed-phase

* Corresponding author: Dipartimento di Chimica, SAPIENZA Università di Roma, Box n° 34 - Roma 62, Piazzale Aldo Moro 5, 00185 Rome, Italy Phone: +39-06-49913679 Fax: +39-06-490631, e-mail: aldo.lagana@uniroma1.it

chromatography. We also discuss the advantages and disadvantages of two different approaches that can be followed for the studies of proteomics: protein-level separation or peptide-level separation.

In: Proteins Researcher Biographical Sketches ...
Editors: H. Z. Wang and M. Tian
ISBN: 978-1-62100-777-7

Chapter 106

Proteomic Approach in Analysing Cardiac Responses on Low-Dose Ionising Radiation Using Cellular and Tissue Models

Soile Tapio*
Helmholtz Zentrum Muenchen – German Research Centre for Environmental Health, Institute of Radiation Biology, Neuherberg

Research Summary

It is well established that high doses of ionising radiation, such as used in radiotherapy, increase risk of cardiovascular diseases (CVD). Observed effects include direct damage to the coronary arteries, marked diffuse fibrotic damage of the pericardium and myocardium, pericardial adhesions, stenosis of the valves and microvascular damage. In contrast, there are considerable uncertainties concerning health effects of low doses of ionising radiation on heart. The need to explore potential biological and physiological effects at low doses is being increasingly acknowledged as the plans for new nuclear power plants and novel medical applications using low-dose radiation are emerging.

The data concerning CVD risk after occupational and environmental exposures to low doses of ionising radiation are controversial. Radiation workers in the Chernobyl liquidator cohort show increased risk for ischemic heart disease. Among employees at British Nuclear Fuels as well as in Canadian nuclear worker cohort and other occupationally radiation-exposed groups there is evidence for an increasing trend concerning circulatory disease mortality with dose. In contrast, no statistically significant increase in circulatory disease mortality due to inhaled radon or external γ-irradiation and its progenies could be observed in German uranium miners. However, while the risk for ischemic heart disease showed no increase, the rate of acute myocardial infarction was enhanced with radon dose.

The most convincing data showing excess radiation-associated risk for CVD has been observed in the Life Span Study of the Japanese atomic bomb survivors. Importantly, even at

* Helmholtz Zentrum Muenchen – German Research Centre for Environmental Health, Institute of Radiation Biology, Ingolstaedter Landstrasse 1, 85764 Neuherberg, soile.tapio@helmholtz-muenchen.de

doses as low as 0.5 Gy the mortality and morbidity due to hypertension and myocardial infarction were increased.

The risk of CVD with low and moderate doses of ionising radiation has recently been reviewed by Little et al..

The vascular endothelium, a continuous monolayer containing of thin, flat cells located on the interior surface of blood vessels, forms an interface between circulating blood and subendothelial matrix. It plays an important role in the integration and modulation of many functions of the arterial wall. Vascular endothelial dysfunction developing during the human aging process seems related to an increased production of reactive oxygen species (ROS). In atherosclerosis, increased endothelial production of ROS leads to oxidation of low density lipoproteins (LDL), accumulation of lipid into foam cells, intimal growth and finally atherosclerotic plaque expansion and rupture.

Whether the biological responses of the endothelium in the case of high and low doses of ionising radiation are similar is still largely unknown. However, in contrast to high-dose radiation, acute doses in the range 0.1–1 Gy result in down-regulation of the adhesion of leukocytes to the endothelium both *in vitro* and *in vivo* and thus may have an anti-inflammatory effect. Furthermore, it is reasonable to believe that not only the dose but also the dose rate has an effect on the biological outcome.

We have used both a human endothelial cell line EA.hy926 and a mouse model to study the immediate proteomic effects of *in vitro* irradiation and long-term functional effects of heart-focussed *in vivo* irradiation, respectively.

As shown in a previous study, EA.hy926 retained most of the characteristics of primary endothelial cells (HUVEC) in a comparative cDNA expression profiling even after addition of statins that are used to reduce the risk of cardiovascular disease. It may thus be considered as a good model system for the cardiac endothelium.

Ea.hy926 was irradiated with 0.2 Gy Co-60 gamma rays with two different dose rates (20 mGy/min and 200 mGy/min) and the cells were harvested 4h and 24 h after the irradiation. The proteome changes in the sham-irradiated vs. irradiated cytosolic fractions were analysed using 2 DE-DIGE techniques. Out of more than fifty protein spots that showed significant alterations in their expression 22 proteins were identified. Among the pathways affected by the low-dose ionising radiation are Ran and Rho/Rock pathways, stress response and glycolysis.

Furthermore, we found across this classification a group of proteins belonging to small Ras-like GTPases, namely Ran, RhoA and Sar1a that share a significant sequence homology, the GDP/GTP binding pocket being especially conserved. Many Ras-superfamily small GTPases are components of signalling pathways that link extracellular signals via transmembrane G-protein-coupled receptors to cytoplasmic or nuclear responses. Interestingly, the previous data show that both RhoA and Ran expression is dependent on the production of reactive oxygen species.

Functional and proteomic alterations in mitochondria isolated from irradiated and sham-irradiated murine (C57Bl6) hearts 4 weeks after heart-focussed irradiation (0.2 Gy, 2 Gy X-ray) were analysed. No differences between sham- and irradiated cardiac mitochondria were found in swelling, respiratory coupling and production of ATP. However, we could identify significantly increased ROS formation in cardiac mitochondria 4 weeks after the exposure to 2 Gy ionising radiation. The results with the irradiated murine hearts emphasize the importance of the persistant oxidative stress as a result of low and moderate doses of ionising radiation. These results are in concordance with a large number of recent data suggesting that altered levels of oxidative stress are essential in the development of cardiovascular disease and that cardiac mitochondria may play an important role both as a target and source of reactive oxygen species.

Taken together, these results demonstrate the importance of proteomics in finding new biological target molecules of a low-dose radiation response in a critical tissue.

In: Proteins Researcher Biographical Sketches ... ISBN: 978-1-62100-777-7
Editors: H. Z. Wang and M. Tian

Chapter 107

SERPIN-RELATED DISEASES

Aleksandra Topic
Institute of Medical Biochemistry, Faculty of Pharmacy, University of Belgrade, Belgrade, Serbia

RESEARCH SUMMARY

Serpins play a critical role in maintaining homeostasis. Approximately two-thirds of human serpins perform extracellular roles, and the rest are localized and function intracellular. There are inhibitory and non-inhibitory serpins. In maintaining homeostasis, serpins are involved in a number of fundamental biological processes such as extracellular matrix remodeling, pro-hormone conversion, intracellular proteolysis, blood pressure, tumor suppression, hormone transport, viral/parasite pathogenicity, modulation of inflammatory response, cell differentiation, cell migration, protein folding, fibrinolysis, complement cascade and blood coagulation.

Severe, human serpin-related diseases are emphysema, liver disease, massive thrombosis or bleeding, hereditary angioedema, dementia, angiopathy and tumor invasion. Inhibitory serpins are 'suicide' or 'single use' inhibitors that use a unique and extensive conformational change to inhibit proteinases. One hallmark of serpins is the reactive centre loop (RCL), a protein motif with a scissile bond, which is cleave by the target proteinase. The structure of the RCL is crucial for the ability of the protein to undergo a 'stressed to relaxed' (S→R) conformational change. The active serpins are in metastable 'stressed form', which is essential for their inhibition role. They can be present in various conformations: native inhibitory with an exposed RCL, latent with a partially inserted RCL or non-inhibitory due to complex formation, cleavage, oxidation of reactive centre residues or polymerization. Any mechanism which reduces the functional level of serpins such as oxidative inactivation, inactivation by non-target proteinases or genetic aberration (mutations), may lead to the pathological process. Serpins are especially vulnerable to mutations, which could alter their native conformations. Aberrations of conformation frequently occur and could lead to a range of diseases that are grouped together and named as the serpinopathies. Nowadays, well-established serpinopathies that reflect the functions and site of synthesis of the individual serpins are emphysema, cirrhosis, thrombosis and dementia. More than 200 different serpins mutations have been identified as the cause of various diseases. Mutations which cause inappropriate conformational change (or misfolding) results in two common outcomes. One is a promotion inappropriate transition to the monomeric latent state with reduction of active inhibitory serpin (antithrombin variants), and the other is serpin polymerization, when the RCL of one serpin molecule inserts into β-sheet A of another to form a dimer, which subsequently extends to form long chains of inactive polymers (alpha-1-antitrypsin variants).

Rarely, mutations in serpin genes cause: severe deficiency with no gene expression (alpha-1-antitrypsin null mutation), dysfunctional variant (alpha-1-antitrypsin PiPittsburgh mutation change function from neutrophil elastase inhibitor into thrombin inhibitor, causing bleeding disorder), and other types. Also, infection with pathogenic organisms which possess a variety mechanism for the interrupting biochemical pathways is an example of inactivation of serpins by non-target proteinases. Namely, pathogen-derived proteinases could cleave RCL and cause broad pathological consequences. Understanding the mechanism of serpin-related diseases opens the perspective to design effective therapy approaches.

In: Proteins Researcher Biographical Sketches ...
Editors: H. Z. Wang and M. Tian
ISBN: 978-1-62100-777-7

Chapter 108

THE ROLES OF MAMMALIAN MITOGEN-ACTIVATED PROTEIN KINASE-ACTIVATING PROTEIN KINASES (MAPKAPKS) IN CELL CYCLE CONTROL

Sergiy Kostenko, Alexey Shiryaev, Nancy Gerits and Ugo Moens*

University of Tromsø, Faculty of Medicine, Institute of Medical Biology, Department of Microbiology and Virology, Tromsø, Norway

RESEARCH SUMMARY

Signal transduction pathways often modulate cell proliferation by targeting the activity of cell cycle regulating proteins. One of the signaling pathways that can control the cell cycle is the mitogen-activated protein kinase (MAPK) signaling cascade. In mammalian cells, the classical MAPK pathway consists of sequential phosphorylation events leading to activation of a MAPK kinase kinase, a MAPK kinase, and a MAPK. MAPK in turn phosphorylates non-protein kinase substrates such as transcription factors, but it can also phosphorylate yet other protein kinases, referred to as MAPK-activating protein kinases (MAPKAPK). Eleven mammalian MAPKAPKs have been identified so far; six of them belong to the group of AGC protein kinases (RSK1, RSK2, RSK3, RSK4, MSK1, and MSK2), while the other five belong to the family of calmodulin-dependent kinases (MK2, MK3, MK5, MNK1, and MNK2). In this review we will discuss those MAPKAPKs that play a role cell-cycle regulation as well as the potential use of specific MAPKAPK inhibitors as therapy in conditions with abnormal cell cycle regulation.

* Corresponding author: Ugo Moens, University of Tromsø, Faculty of Medicine, Institute of Medical Biology, Department of Microbiology and Virology, N-9037 Tromsø, Norway. Phone: +47-77644622, fax: +47-77645350, e-mail: ugom@fagmed.uit.no

In: Proteins Researcher Biographical Sketches ... ISBN: 978-1-62100-777-7
Editors: H. Z. Wang and M. Tian

Chapter 109

RHO-KINASE INHIBITOR IN KIDNEY DISEASE

Toshio Nishikimi*
Department of Hypertension and Cardiorenal Medicine,
Dokkyo Medical University, Mibu, Tochigi, Japan

RESEARCH SUMMARY

Recent studies demonstrated the importance of small GTP binding protein in physiological function. Rho, a member of small GTP binding protein, is known to function as a molecular switch in various cellular functions, including contraction, actin cytoskeleton organization, cell adhesion and motility, proliferation, cytokinesis, and gene expression. Among the Rho effectors, the cellular function and signal transduction of Rho-kinase have been extensively studied. However, the information about in vivo function is still limited until a specific inhibitor of Rho-kinase such as Y-27632 and fasudil was discovered. Rho-kinase inhibitor is a very powerful tool for examining the role of the Rho/Rho-kinase pathway in vitro and in vivo. Recent studies have demonstrated that Rho/Rho-kinase pathway is involved in the pathogenesis of various cardiovascular diseases such as arteriosclerosis, hypertension, angina pectoris, myocardial infarction, and pulmonary hypertension. However, there are few studies, which investigated the role of Rho/Rho-kinase in renal disease. Here we review the information about the therapeutic importance of the Rho/Rho-kinase pathway in renal disease. Specifically, we describe the recent our results about beneficial effect of Rho-kinase inhibitor on the glomerulosclerosis in hypertensive nephropathy. Our results suggest that chronic inhibition of Rho-kinase pathway may be a new therapeutic approach for the hypertensive glomerulosclerosis. Our results also suggest that the possible mechanism of renoprotective effect of Rho-kinase inhibitor is mediated via many pathways, including inhibition of extracellular matrix gene expression, monocytes/macrophages infiltration, and oxidative stress, and upregulation of eNOS gene expression.

* Correspondence author: TEL: 81-282-87-2149, FAX: 81-282-86-1596, e-mail: nishikim@dokkyomed.ac.jp

In: Proteins Researcher Biographical Sketches ... ISBN: 978-1-62100-777-7
Editors: H. Z. Wang and M. Tian

Chapter 110

TARGETING THE EPIDERMAL GROWTH FACTOR RECEPTOR PATHWAY IN GLIOBLASTOMA MULTIFORME AND OTHER INTRACRANIAL MALIGNANCIES

Marc-Eric Halatsch*[*] and Georg Karpel-Massler
Department of Neurosurgery, University of Ulm School of Medicine, Steinhövelstraße, Ulm, Germany

RESEARCH SUMMARY

Limited therapeutic options exist for glioblastoma multiforme (GBM), the most common primary central nervous system tumor, and recurrence is common. Standard therapy is surgical resection, where possible, and radiotherapy. Adjuvant chemotherapy provides only a modest survival benefit, and new therapies are urgently needed. Dysregulated epidermal growth factor receptor (HER1/EGFR) is found in 40-50% of GBM. As the intracellular tyrosine kinase (TK) of the HER1/EGFR activates signalling cascades leading to cell proliferation, angiogenesis and inhibition of apoptosis, HER1/EGFR represents an attractive therapeutic target. HER1/EGFR TK inhibitors are in advanced clinical development for glioma, and a number of trials are in progress, or have recently been completed. Although data from experimental studies seem promising, proof of a significant clinical benefit is lacking. A problem that has to be further addressed is the prediction of the individual tumor response to HER1/EGFR TK inhibitors based on molecular determinants. This chapter reviews the role of HER1/EGFR TK inhibitors in the treatment of GBM and other intracranial malignancies.

* Corresponding author: Tel: +49 731 50055003, Fax: +49 731 50055002, email: marc-eric.halatsch@uniklinik-ulm.de

In: Proteins Researcher Biographical Sketches ...
Editors: H. Z. Wang and M. Tian

ISBN: 978-1-62100-777-7

Chapter 111

THE SERINE PROTEINASE INHIBITOR Z ALPHA-1 ANTITRYPSIN: ACTING ON THE NF-KAPPAB SYSTEM FOR CYTOTOXICITY

Matthew William Lawless*

Department of Clinical Medicine, Trinity College Dublin,
St. James Hospital, Dublin 8, Ireland

RESEARCH SUMMARY

The serine proteinase inhibitors Alpha-1 Antitrypsin (A1AT) and its mutant protein the so-called 'Z' mutation in A1AT deficiency encodes a glutamic acid-to-lysine substitution at position 342 in A1AT and is the most common A1AT allele associated with disease. In recent years, several important research advances in the cell biology of this aggregation-prone serine proteinase inhibitors and its pathobiological mechanisms for disease onset have proven paramount to our understanding of Z A1AT deficiency. The liver disease associated with Z A1AT, occurs in up to 15% of A1AT-deficient individuals, a result of toxic gain-of-function mutations in the A1AT gene, which cause the A1AT protein to fold aberrantly and accumulate in the endoplasmic reticulum of hepatocytes. The lung disease is associated with loss-of-function, specifically decreased anti-protease protection on the airway epithelial surface. Most strikingly, this polymer has now been revealed to activate the NF-kappaB (NFκB) system and cause cytotoxicity by a pathway that is independent of the unfolded protein response (UPR). NFκB which is an inducible transcription factor which regulates the expression of a range of genes involved in important biological processes such as the innate and adaptive immunity, inflammation, cellular stress responses, cell adhesion, apoptosis and proliferation. This chapter addresses the structural basis of this serine proteinase inhibitor (Z A1AT) and how the ordered accumulation of this polymer activates the NFκB system

* Corresponding author: Email: mlawles@tcd.ie, lawlessm@hotmail.com

affecting cell death. Furthermore, we place these findings in the context of the histone code in order shed light on how to target the NFκB system in a specific fashion for the setting of Z A1AT disease.

In: Proteins Researcher Biographical Sketches … ISBN: 978-1-62100-777-7
Editors: H. Z. Wang and M. Tian

Chapter 112

SRC FAMILY KINASE INHIBITORS IN CANCER THERAPY

***Faye M. Johnson*[*1] *and Gary E. Gallick*[2]**

[1] Department of Thoracic/Head and Neck Medical Oncology,
The University of Texas M. D. Anderson Cancer Center,
and The University of Texas Graduate School of Biomedical Sciences at Houston,
Houston, TX, US

[2] Department of Cancer Biology,
The University of Texas M. D. Anderson Cancer Center,
and The University of Texas Graduate School of Biomedical Sciences at Houston,
Houston, TX, US

RESEARCH SUMMARY

The first cellular oncoprotein to be studied was the viral protein v-src, a tyrosine kinase. Protein tyrosine kinases were found later to be key regulators of cellular signaling pathways that control important processes in cancer progression, including transformation, proliferation, survival, invasion, and angiogenesis. Protein tyrosine kinase inhibitors are a relatively new class of agents that have been developed to exploit the importance of tyrosine kinases in cancer biology. One potential therapeutic target for which agents have been developed recently is the src family of kinases (SFKs). SFKs are nonreceptor tyrosine kinases involved in signal transduction in a wide variety of malignancies. Interest in SFKs has increased because of the development of pharmacologic SFK inhibitors that have exhibited initial clinical success and low toxicity. c-Src is the best-studied member of the Src family and the one most often involved in cancer progression. c-Src has multiple substrates, and its inhibition can lead to changes in proliferation, motility, invasion, survival, and angiogenesis.

[*] Address correspondence to Faye M. Johnson at the Department of Thoracic/Head and Neck Medical Oncology, Unit 432, The University of Texas M. D. Anderson Cancer Center, 1515 Holcombe Boulevard, Houston, TX 77030, USA; Tel: 713-792-6363; Fax: 713-792-1220; E-mail: fmjohns@mdanderson.org

SFKs have been demonstrated to be important for the progression of many epithelial tumors, such as cancers of the colon, pancreas, breast, lung, head and neck, and prostate. SFKs also are expressed in sarcomas, melanoma, and hematologic malignancies. Several inhibitors of the SFKs are in clinical development. Four SFK inhibitors are currently being studied in clinical trials and have been well tolerated; data on their antitumor effects are pending. Future clinical development of these inhibitors will include completion of trials in patients with solid tumors, definition of sensitive populations, and trials of combination therapy.

In: Proteins Researcher Biographical Sketches ...
Editors: H. Z. Wang and M. Tian
ISBN: 978-1-62100-777-7

Chapter 113

PROTEIN KINASE INHIBITORS IN CANCER

Yiguo Hu and Shaoguang Li*
The Jackson Laboratory, Bar Harbor, Maine, US

RESEARCH SUMMARY

A protein kinase is an enzyme that modifies other proteins by chemically adding phosphate groups to them (phosphorylation). Phosphorylation usually results in a functional change of the target protein (substrate) by changing enzyme activity, cellular location, or association with other proteins. Protein tyrosine kinases (PTKs) play a key role in the regulation of cell proliferation, differentiation, metabolism, migration, and survival. Due to their involvement in various forms of cancers, PTKs have become prominent targets for therapy. There are two principles to developing PTK inhibitors. As binding APT is essential for kinase activity, the common strategy is to look for the small molecules that can compete the ATP binding sites. To increase specificity of PTK inhibitors, a more attractive strategy is to develop non-ATP-competitive kiniase inhibitors. So far, there are many PTK inhibitors that have been developed. Several inhibitors have been successfully used to treat human cancers in clinic. These agents are shown to inhibit multiple functions of cancer cells, including proliferation, survival, invasion, and angiogenesis.

* Corresponding author: Tel: 207-288-6734, Email: shaoguang.li@jax.org

In: Proteins Researcher Biographical Sketches ...
Editors: H. Z. Wang and M. Tian
ISBN: 978-1-62100-777-7

Chapter 114

PROTEIN KINASE INHIBITORS IN THE TREATMENT OF MALIGNANT LIVER AND KIDNEY TUMORS

Panagiotis Samaras and Frank Stenner*
Department of Oncology, University Hospital Zurich,
Raemistrasse, Zurich, Switzerland

RESEARCH SUMMARY

Treatment options for cancer have increased in the last years due to the ongoing research and development of new therapeutics. A focus shift from formerly undirected cytostatic agents to more targeted therapies, mainly antibodies and protein kinase inhibitors has taken place. New agents directed against receptors which control proliferation, migration and angiogenesis of tumor cells have been developed. Mainly specific antibodies directed against the extracellular ligand binding domain of various receptor tyrosine kinases, and protein kinase inhibitors targeting the intracellular protein kinase domain of receptors or cytoplasmic signalling proteins have been increasingly developed in the past years. In this chapter we will address the current treatment options with protein kinase inhibitors exemplarily for hepatocellular carcinoma and renal cell cancer. We will further give an insight into the actual research perspectives and the conception of incorporating new drugs into clinical trials as single agent treatment or combined with "classical" chemotherapy. The first part elucidates the molecular pathways of carcinogenesis and depicts the mode of action of these new agents in the treatment of hepatocellular carcinoma, the second part illustrates the pathogenesis of renal cell cancer and the resulting therapeutic options by targeting these pathways.

* Corresponding author: Phone: +41 (0) 44 255 2214 Fax: +41 (0)44 255 45 48 frank.stenner@usz.ch

In: Proteins Researcher Biographical Sketches ...
Editors: H. Z. Wang and M. Tian
ISBN: 978-1-62100-777-7

Chapter 115

PAI-1 AND THE DIET-INDUCED OBESITY PHENOTYPE: BACKGROUND EFFECTS AND INBREEDING

Bart M. De Taeye*, Tatiana Novitskaya and Douglas E. Vaughan
Department of Medicine, Division of Cardiovascular Medicine,
Vanderbilt University, Nashville, TN, US

RESEARCH SUMMARY

The effect of PAI-1 deficiency on the development of obesity and insulin resistance is controversial. Published reports have suggested that PAI-1 deficiency can protect or have no effect on the development of diet-induced obesity in mice. To further investigate these apparent discrepant findings, PAI-1 deficient and wild-type C57BL/6J littermate mice were fed a diet high in fat or a regular diet and this for either 12 or 30 weeks. The development of obesity and the degree of insulin resistance was analyzed. PAI-1 deficient animals were not protected from the development of obesity or insulin resistance. Analysis of the change in weight and percentage body fat for the individual mice revealed that even though mice were littermates, they showed significant variability in the degree of obesity development and adipose tissue accumulation within the same group, in agreement with findings from studies focusing on the effects of epigenetic changes in diet-induced obesity in littermate mice. As such, while our data confirm that PAI-1 genotype does not influence the development of diet-induced obesity and diabetes, they help explaining so far contradicting data. In conclusion, PAI-1 inhibition could be important to prevent complications associated with obesity but not the development of obesity itself.

* Corresponding author: Vanderbilt University – Medical Center, Department of Medicine, Division of Cardiovascular Medicine, 352 Preston Research Building, Nashville, TN 37232, USA, Phone: (615) 936-1719, Fax: (615) 936-2936, E-Mail: Bart.detaeye@vanderbilt.edu

In: Proteins Researcher Biographical Sketches ...
Editors: H. Z. Wang and M. Tian

ISBN: 978-1-62100-777-7

Chapter 116

THE STRUCTURE OF α_1-PROTEINASE INHIBITOR POLYMER: FACTS AND HYPOTHESES

Ewa Marszal
Division of Hematology, Office of Blood Research and Review,
Center for Biologics Evaluation and Research, Food and Drug Administration,
Bethesda, MD, US

RESEARCH SUMMARY

The metastable structure of a serpin α_1-proteinase inhibitor (α_1-PI) makes it prone to conformational changes as a result of certain mutations and under mild denaturing conditions. Polymers of several variants of α_1-PI, e.g., the most clinically relevant Z variant, accumulate in the liver and are believed to be the underlying cause of the liver disease associated with α_1-PI deficiency. Such polymers have also been found in the circulation and in the lung. The structure of these polymers, which is important for structure-based drug design, remains unknown. The historically first and generally accepted model of the α_1-PI polymers, the loop-A sheet model, which assumes insertion of the reactive center loop (RCL) of one α_1-PI molecule between the central strands of the A β-sheet of another molecule, has never been proven. In addition, this model is inconsistent with resonance energy transfer data obtained for heteropolymers formed from the Z and S variants. It is also difficult to envision how this model or even the more compact loop-A sheet model, deduced from fluorescence data obtained for the Z and S variant heteropolymers, can be compatible with electron microscopy (EM) images, which show apparently flexible polymer chains. Two other polymer models were deduced from the interactions seen between molecules in the crystals of two other serpins, antithrombin (AT) and plasminogen activator inhibitor-1 (PAI-1), which demonstrated the ability of the reactive center loop (RCL) to assume the β-strand conformation and to extend the C or A β-sheet in a neighboring molecule by formation of an additional edge strand. This became the ground for the loop-C sheet and strand s7A models of the polymer, although the interactions seen in crystals appear not to be stable in solution and, thus, appear not to reflect interactions existing in α_1-PI polymers, which actually are stable.

The focus of this paper is to review the observations used to support the proposed models and to revisit the wealth of literature data in search for unexplained observations and possible new interpretations. I also put forward the hypothesis that a fusion of β-sheets, rather than insertion of the reactive center loop into a β-sheet, may underlie the polymerization process of α_1-PI. This hypothesis is an expansion of the recent head-to-head model proposed based on the polymerization of the α_1-PI disulfide-linked dimer.

In: Proteins Researcher Biographical Sketches … ISBN: 978-1-62100-777-7
Editors: H. Z. Wang and M. Tian

Chapter 117

SERPINA5 Expression in the Male Reproductive Tract Is Altered with Advanced Age

Matthew D. Anway
Center for Reproductive Biology; Department of Biological Sciences, University of Idaho; Moscow, ID, US

Research Summary

The serine proteinase inhibitor SERPIN5A is abundantly expressed in the male reproductive tract in rodents and humans. SERPIN5A is critical for maintaining proper development of sperm. In mouse model, SERPIN5A knockout males are infertile. SERPIN5A is responsible for inactivating serine proteinases, such as protein c, urokinase and tissue plasminogen activator. SERPIN5A has been associated with several disease states including infertility, thrombosis and prostate disease in humans and rodents. The expression and localization of SERPIN5A is well documented in the early development of the male reproductive tract. However, the mRNA expression pattern of SERPIN5A in the aging male reproductive tract is unclear. The Brown Norway rat develops reproductive tract abnormalities with advanced age. To this end, we investigated the transcript level of SERPIN5A in the young (4 month), mid age (10 month) and aged (19 month) male reproductive tract. SERPIN5A was differentially expressed in the male reproductive tract with advanced age. The gene expression of SERPIN5A was up-regulated in the Sertoli cells, dorsal and ventral prostate tissue with advanced age. The epididymis had the largest increase between the young and the aged tissue. SERPIN5A was down-regulated in the lateral prostate tissue. SERPIN5A had the highest expression in the seminal vesicles, but was not differentially regulated with age. The expression of SERPIN5A is sensitive to androgens and although testosterone levels are lower with advanced age it is not responsible for the observed alterations in the gene expression. We hypothesize that the changes in SERPIN5A expression in the male reproductive tract with advance age in the Brown Norway rat is the result of the disease progression.

In: Proteins Researcher Biographical Sketches ...
Editors: H. Z. Wang and M. Tian
ISBN: 978-1-62100-777-7

Chapter 118

VASPIN: VISCERAL ADIPOSE TISSUE-DERIVED SERPIN WITH INSULIN-SENSITIZING EFFECTS

Jun Wada*
Department of Medicine and Clinical Science,
Okayama University Graduate School of Medicine,
Shikata-cho, Okayama, Japan

RESEARCH SUMMARY

Metabolic syndrome is characterized by abdominal obesity and clustered with atherogenic risk factors, *i.e.,* hypertension, dyslipidemia, hyperglycemia, and hyper insulinemic insulin resistance. As a consequence, morbidity and mortality of cardiovascular disease and stroke markedly increased in the subjects with metabolic syndrome. Vigorous efforts were made to delineate the molecular link between increased adiposity and insulin resistance and it demonstrated that a variety of substances, including free fatty acids, leptin[1], tumor necrosis factor-α (TNF-α), acylation-stimulating protein (ASP), adiponectin, and resistin, are secreted from adipose tissues and they modulate the insulin sensitivity. In addition to insulin sensitivity, several adipocytokines directly affect the vascular functions. For instance, the investigation using globular domain adiponectin (gAd) transgenic (Tg) mice revealed that globular adiponectin can protect against atherosclerosis *in vivo*[6]. Thus, the decreased levels of adiponectin in obese subjects may attribute to the development of atherosclerosis in patients with metabolic syndrome. It has been recognized that relevance of different adipose tissue depots to insulin resistance, *i.e.* the accumulation of visceral adipose tissues such as mesenteric fat depots, closely relates with insulin resistance and development of metabolic syndrome. The discovery of various adipokines also supported this notion and this basic concept was introduced to the IDF (International diabetes foundation) consensus worldwide definition of metabolic syndrome in 2005, in which the subjects with metabolic syndrome must have central obesity defined by waist circumference (http://www.idf.org/webdata/docs/IDF_Meta_def_final.pdf).

---

*Corresponding author: Phone: +81-86-235-7235, Fax : +81-86-222-5214, E-mail: junwada@md.okayama-u.ac.jp

In visceral adipose tissues in genetically obese rats, *i.e.* Otsuka Long-Evans Tokushima fatty (OLETF) rats, we have found that some serine protease inhibitors (serpins) are differentially expressed during the development of obesity revealed by DNA microarray and PCR-based cDNA subtraction methods. During investigation of such serpin genes, we identified that Serpina12 is specifically up-regulated in visceral adipose tissues of OLETF rats. Thus, we designated a new serpin member as visceral adipose tissue-derived serine protease inhibitor, vaspin. In this commentary, I would like to describe the functional link between vaspin and metabolic syndrome and the future perspectives for the drug discovery and development related to vaspin

In: Proteins Researcher Biographical Sketches ...
Editors: H. Z. Wang and M. Tian

ISBN: 978-1-62100-777-7

Chapter 119

EFFECT OF ALPHA2-ANTIPLASMIN ON TISSUE REMODELING

Yosuke Kanno* and Hiroyuki Matsuno
Department of Clinical Pathological Biochemistry, Faculty of Pharmaceutical Science, Doshisha Women's College of Liberal Arts, Kodo Kyo-tanabe Kyoto, Japan

RESEARCH SUMMARY

The fibrinolytic system (Plasminogen/plasmin system) is supposed to play an important role in both the synthesis and degradation of extracellular matrices. However, the detailed mechanism on how this system affects tissue remodeling remains unclear. Inhibition of the system occurs either at the levels of plasminogen activator, regulated by specific plasminogen activator inhibitors (PAIs) or at the levels of plasmin, mainly regulated by alpha2-antiplasmin (α2AP). α2AP is a specific plasmin inhibitor. We investigated the role of α2AP on tissue remodeling by using a wound-healing model and fibrosis model in both wild type mice and α2AP deficient ($\alpha 2AP^{-/-}$) mice. Our findings newly indicate that the absence of α2AP enhances the secretion of VEGF, and over secretion of VEGF promotes angiogenesis and wound healing. Moreover, the absence of α2AP attenuates bleomycin-induced fibrosis, and α2AP induces the production of TGF-β. These data suggest that α2AP plays an important role in wound healing and fibrosis. These findings indicate a potential new aspect in this field and could be a useful report on tissue remodeling.

* Corresponding author: Phone: +81(Japan) 0774-65-8629, E-mail: ykanno@dwc.doshisha.ac.jp